PRAISE FO
MID-COURSE CORRECT

"Unlike most business leaders for whom 'the business case for sustainability' is all that really matters, Ray Anderson unapologetically advanced a moral case as well, constantly focused on our duty to future generations. This is more important today than ever before, as we come to recognize that an incremental, softly-softly approach to corporate sustainability is pretty much a busted flush—we've simply run out of time. The Interface story is as inspirational today as ever, but it needs to be read for its deeper, radical reckoning: If not now, when? If not you, who?"

—JONATHON PORRITT, founder and director,
Forum for the Future; author of *The World We Made*

"I'm so glad Ray Anderson's story is getting another telling—few sagas are more inspiring or more timely. We desperately need more and more people following in his footsteps with the same blend of humility and determination!"

—BILL MCKIBBEN, author of *Falter*

"Twenty years after its first edition, there is still so much for us to harvest and learn from *Mid-Course Correction*. When it came to the precariousness of our shared future, Ray Anderson was both impatient and relentless in fighting for a world of beauty, abundance, justice, and fairness. When Ray asked me to join the Interface board, his exact words were, 'Come help me change the world!' Those words stayed with me throughout my seventeen years working with him. This twenty-year update provides the perfect guide for others to join in climbing Mt. Sustainability, the most critical mission of our time."

—DIANNE DILLON-RIDGLEY, CEO,
Women's Network for a Sustainable Future

"So far, Ray C. Anderson is the twenty-first century's undisputed master of making business a potent force for saving people and the planet. As his winning carpet and textile firm, Interface, now wrings out the last few percent of its fossil-fuel use, his bold strategy—take nothing, waste nothing, do no harm, do very well by doing good—inspires visionary leaders

everywhere. This valuable update, with additions from his grandson, John Lanier, maps out necessary next steps."
—**AMORY B. LOVINS**, cofounder and chief scientist, Rocky Mountain Institute; author of *Reinventing Fire*

"Twenty-one years ago my friend Ray Anderson brought an engineer's insight, a businessman's rigor, a grandfather's love, and a poet's heart to what he called 'the creative act of business.' He challenged his company to 'first to attain sustainability and then to become restorative,' reminding all who would listen that 'if your sustainability program is costing you money, you're doing it wrong.' And in this book and in his countless speeches—with a vision as clear as any since, to our peril and shame, and with a roadmap still valid—he challenged us all to do the same."
—**GIL FRIEND**, CEO, Natural Logic, Inc.; founder, Critical Path Capital

"Ray Anderson was one of the most extraordinary business leaders I ever met—and I have met and worked with scores. He was extraordinary in his early embrace of the sustainability agenda, years before most of his peers were even aware of the term. And he was extraordinary in his willingness to admit he had got parts of his response wrong, which is the remarkable tale brought bang up to date in *Mid-Course Correction Revisited*. Highly recommended for anyone wanting leadership in these challenging times."
—**JOHN ELKINGTON**, founder and chief pollinator, Volans; originator of the Triple Bottom Line

"When I began my personal journey from a traditional business career to this world of 'sustainability,' Ray Anderson's *Mid-Course Correction* was the first book I read. I felt the same 'spear in the chest' that Ray described, and so I followed his intellectual path of discovery. I am indebted to Ray's legacy, and I know it is long past time to revisit his work. The global challenges we face are more daunting than ever, so the imperative Ray described has only gotten more urgent. We must convert 'business as usual' from an obsession with short-term profits to a relentless focus on using business to build a thriving world. Ray saw it clearly years before almost everyone, and it's a critical time to bring his vision to a new generation of business leaders."
—**ANDREW WINSTON**, founder, Winston Eco-Strategies; author of *The Big Pivot* and coauthor of *Green to Gold*

Mid-Course Correction

REVISITED

The Story and Legacy of a Radical Industrialist
and His Quest for Authentic Change

Ray C. Anderson
and John A. Lanier

Foreword by Paul Hawken

Chelsea Green Publishing
White River Junction, Vermont
London, UK

This book is an expansion of the original *Mid-Course Correction*, published in 1998
by its author, Ray C. Anderson.

Mid-Course Correction Revisited was published in collaboration with the Ray C. Anderson Foundation, which
works to advance Ray Anderson's legacy and vision of business and industry helping to create a sustainable
world for future generations.

Editor: Joni Praded
Project Manager: Patricia Stone
Copy Editor: Laura Jorstad
Proofreader: Angela Boyle
Indexer: Shana Milkie
Designer: Melissa Jacobson

Printed in Canada.
First printing April 2019.
10 9 8 7 6 5 4 3 2 1   19 20 21 22 23

Our Commitment to Green Publishing

Chelsea Green sees publishing as a tool for cultural change and ecological stewardship. We strive to align our
book manufacturing practices with our editorial mission and to reduce the impact of our business enterprise
in the environment. We print our books and catalogs on chlorine-free recycled paper, using vegetable-based
inks whenever possible. This book may cost slightly more because it was printed on paper that contains recycled fiber, and we hope you'll agree that it's worth it. Chelsea Green is a member of the Green Press Initiative
(www.greenpressinitiative.org), a nonprofit coalition of publishers, manufacturers, and authors working to
protect the world's endangered forests and conserve natural resources. *Mid-Course Correction Revisited* was
printed on paper supplied by Marquis that is made of recycled materials and other controlled sources.

Library of Congress Cataloging-in-Publication Data
Names: Anderson, Ray C., author. | Lanier, John A., author.
Title: Mid-course correction revisited : the story and legacy of a radical industrialist and his quest for
    authentic change / Ray C. Anderson and John A. Lanier ; foreword by Paul Hawken.
Other titles: Mid-course correction
Description: White River Junction, Vermont : Chelsea Green Publishing, 2019. | Revised edition,
    includes new foreword by Paul Hawken and several new chapters by John A. Lanier. | Includes
    bibliographical references and index.
Identifiers: LCCN 2018059113| ISBN 9781603588898 (pbk.) | ISBN 9781603588904 (ebook)
Subjects: LCSH: Interface, Inc. | Sustainable development—United States—Case studies. | Social
    responsibility of business—United States—Case studies. | Industrial management—Environmental
    aspects—United States—case studies.
Classification: LCC HC110.E5 A6616 2019 | DDC 658.4/08—dc23
LC record available at https://lccn.loc.gov/2018059113

Chelsea Green Publishing
85 North Main Street, Suite 120
White River Junction, VT 05001
(802) 295-6300
www.chelseagreen.com

MIX
Paper from
responsible sources
FSC® C103567

To environmentalists everywhere, from the giants of the environmental movement to the ordinary folks who got it long before I did, I dedicate this little book. Thank you all for the inspiration. Would that I could, in the time I have left as a visitor to this beautiful blue planet, the third from a star called "Sol" in a galaxy called "Milky Way," do justice to your lives and examples with my own.

—RCA
July 1998

For Bailey, Brooks, Banks, J.R., August, and Cecilia, and for any of Ray's great-grandchildren yet to walk this beautiful, blue-green living planet.
Each of you is Tomorrow's Child.
He did all of this for you.

—JAL
2018

Contents

Foreword

Those who were fortunate to know Ray Anderson were honored, delighted, and sometimes in awe. He was many people—father, executive, colleague, brother, speaker, writer, leader, pioneer—but I am not sure any of us ever quite figured him out. On the outside, Ray was deceptively traditional, and sometimes very quiet. He was an everyman— an all-American, down-home Georgian. He was so seemingly normal that he could say just about anything and get away with it because people didn't quite believe what they had heard. The metaphorical *spear in his chest* that Ray writes about describes an awakening that led him to give talks all over the world and become a great teacher. He could walk into an audience and leave listeners transfixed by a tenderness and introspection they never expected and had never encountered before. He was also courageous. He stood in front of large groups of learned individuals telling them that pretty much everything they were doing was destroying the earth. And then he followed that statement with a personal assessment: He, too, was a thief and robber stealing the future of civilization because of the way he practiced commerce. He meant every word, and his message landed deeply in the hearts and minds of hundreds of thousands of people. Business audiences in particular had no defenses because they had no framework for Ray. Was he really a businessman? Yes. Was he a conservative southern gentleman with a refined drawl? Yes. Was he successful? For sure. Where, then, did these radical statements come from? Paradoxically, because people could not connect the dots, he was extraordinarily credible.

In *Mid-Course Correction Revisited,* you will find the pragmatist. Because Ray's goals went far beyond the environmental commitments of every corporation in the world, he was called a dreamer. And to be sure, he was. But he was also an engineer. He had definitely seen the mountain, but he also dreamed in balance sheets, thermodynamics, and resource flow theory. He dreamed a world yet to come because dreams of a livable future do not come from politicians, bankers, and the media. For Ray, reimagining the world was a responsibility, something owed to our children's children, a gift to a future that is begging for selflessness and vision. And here we read about how he set his mind and life to that task.

He used business as a means to educate and transform, but his life was not about money or carpets. Ray's life was about the sacred. His covenant was with God; the marketplace is where he labored. His teachings are a lineage that will live on for centuries to come. He realized that all commerce depends upon a sacred Earth with all its complexity, beauty, and mystery. A newborn awareness of the interdependence and connectedness of all life was Ray's epiphany. Its attendant responsibility to do no harm was his mission. He was free to reimagine the relationship between humanity and nature with Interface as the model. There were no longer human systems and ecosystems; there was one system, and he understood that the laws of physics and biology prevailed. He believed in Emerson's words that there is an innate morality in the laws of nature: "I have confidence in the laws of morals as of botany. I have planted maize in my field every June for seventeen years and I never knew it come up strychnine. My parsley, beet, turnip, carrot, buck-thorn, chestnut, acorn, are as sure. I believe that justice produces justice, and injustice injustice." Ultimately Ray's work was not about making a sustainable business; it was about justice, ethics, and honoring creation. His goal of zero waste was the path to 100 percent respect for all living beings.

Having said all this, we don't know exactly what happened to Ray when he had his epiphany in 1994. Yes, he had read my book, *The Ecology of Commerce*. However, something remarkable was already within his being, and it came to life. It is as if from that point twenty-four years ago, Ray could truly see. He saw benevolence and beauty, the tightly knit longleaf pine forests, the undulant riverine corridors of the Chattahoochee, the tantalizing pure light reflected on bracts and fronds, the drifting silvery spider silk that takes tiny passengers to new forests. Once your eyes open to the magnificence of creation, you cannot unsee it. And Ray never looked back. He did not ponder long. He went to work. He was not satisfied by simply being able to see; he was destined to serve life itself, for what else is there to do once you realize how phenomenally we are stitched together by the living world? He did not see nature as an abstraction to be worshiped but as the matrix of reformation, the source of goodness, the architecture of our spirit, and the template of a future delineated by people who know that business has no purpose if not to serve and honor all of life. He saw that our lives rely upon the kindness of strangers and the damp forest floor and spirited grasses and the human community. His life is a testament to that love.

Ray's physical presence has vanished into a mystery we will all follow but never fully understand. His dream, his yearning for commerce that regenerates life and does no harm, and his intention to reconceive what it means to be a manufacturer and to bring industry and biology together into one entity, burned in him a flame that never ceased, and it lives on in his company and thousands of others. For me he was the greatest leader in industrial ecology, the businessman who defined and demonstrated how commerce must be if we are to continue our life here on Earth. We are blessed by his presence, thoughts, and deeds. Proverbs reminds us that though all good people die, goodness does not perish. Herein is a roadmap of the goodness Ray created and offered to the world.

—PAUL HAWKEN

PART ONE

Mid-Course Correction

by Ray C. Anderson

A Note from John Lanier

In the pages that follow, you will read the original *Mid-Course Correction*, recounting the now classic journey of Ray C. Anderson, a carpet manufacturer whose spear-in-the-chest moment woke him up to environmental issues, caused him to rethink his entire way of doing business, and turned him into a green business pioneer whose commitment and vision remain in a class of their own. These are the words with which Ray laid out his vision for the "prototypical company of the twenty-first century."

Twenty years after its first publication, Ray's story remains an inspiration. Green business leaders still ask, *Why aren't there more Ray Andersons?* The company he founded, Interface, has become a global leader in environmental stewardship, constantly pushing the boundaries of what corporate sustainability means. Which routinely prompts another question: *Why aren't there more Interfaces?* In part 2 we will visit the many people at work to make those questions obsolete, learn how Interface transformed itself over the last two decades, and explore sustainability's most difficult frontier, the transformation of commerce and the creation of the prototypical *economy* of the twenty-first century.

For now, turn the clock back to 1998 and experience the making of a radical industrialist.

Please note that some updates have been made to the original text of *Mid-Course Correction* for accuracy.

Prologue

Years ago there was a popular television show titled *I Led Three Lives*. It was about the adventures of a businessman and double agent who lived his three lives simultaneously. Well, I, too, have lived three lives, but I have lived them sequentially and am still in my third. (I mean lives from a professional standpoint.)

My first life spanned the first thirty-eight years of my sojourn on Earth, during which I prepared myself, in countless ways, to be the entrepreneur who would found the company that came to be called Interface, Inc.

My second life began with that act of creation—founding a successful company is as creative an act as there is, including the requisite 99 percent perspiration—and went for another twenty-one years. During that time, Interface, Inc., founded to produce and sell carpet tiles for American office buildings, survived start-up and prospered beyond anyone's dreams. But twice in its twenty-five years of existence, it has hit the wall. Once was in 1984 when its primary market—new office construction in the United States—collapsed. That time the company reinvented and re-formed itself quickly by diversifying its marketplace to include office building renovations, other market segments such as health care facilities, and other geographic markets; and by entering other businesses, specifically textiles and chemicals, mostly by acquisitions that were financed through public offerings of equity and debt.

Interface hit the wall again during 1991–93 because of several more or less coincident occurrences: the worldwide movement to downsize corporations, which dampened a key market segment for the company (companies tend not to buy carpet when they are laying off people); worldwide recession; new and tougher competition; and market preferences shifting toward less expensive products that we were not very good at making. The company, once again, reinvented itself, but this time with the infusion of a new management team, led by Charlie Eitel, that brought new ideas and new energy. Thus, this second reinvention was more profound than the first, and completely changed my life as founder, chairman, and CEO of the billion-dollar company that Interface had become.

The new management team took hold of operations quickly and effectively, and my job became one of turning loose, getting out of the way, staying out of the way, and being head cheerleader. That's a big change after twenty-one years of nose-to-the-grindstone, autocratic, hands-on management. I began seriously to question my role, what it should be, and if indeed I had one.

Then, in a sequence of events that I describe in the pages that follow, I discovered an urgent calling and an unexpectedly rewarding new role for myself. Thus began my third life, with a new vision of what I wanted Interface, my child, to grow up to be.

One aspect of my new role is being spokesperson to a growing audience that is hungry to hear the Interface story. Trying to satisfy that hunger, I make a lot of public speeches. This little book expresses the beliefs I have embraced and the convictions I have formed as I have prepared those many speeches, all about this urgent calling I refer to as my third life and the vision it has produced for my company and, I fervently hope, will foster for others as well.

CHAPTER ONE

The Next Industrial Revolution

I like to open my speeches, especially those that deal with the environment, by doing something I learned from my friend, the outstanding family therapist and popular speaker J. Zink. I like to have the audience stand and have each person pick out a couple of people standing nearby, preferably strangers, and give them a hug. It never fails to create a lighthearted bond between the audience and me, and it gives me the basis for making a strong opening statement: "I hope the symbolism of that was not lost on you, fellow astronauts on Spaceship Earth. We have only one spaceship. It's in trouble. We're in this together and need each other." According to Dr. Zink, the hug also opens up access to the right hemispheres of people's brains where he says feelings, including conscience, reside. Then I like to say that, despite the initial levity, my assignment (self-proclaimed) is to disturb, not amuse; to inform, not entertain; and to sensitize (or further sensitize) my audience to the crisis of our times and of all time to come. I invite my audience, if they find me radical and provocative, to be provoked to radical new thinking, and I suggest that all of us need to do more of that.

On a Thursday in April 1996, I was in Boston on a panel speaking to five hundred people. The subject was "Planning for Tomorrow," and the panel was about technology's role and impact on the strategic decisions companies make. The discussion was sponsored by the International Interior Design Association. The audience was about one-third interior designers and two-thirds businesspeople, including some of my company's competitors.

While the subject of the discussion was technology, I think that the audience's understanding of the term probably had to do with the technology in the offices where most of them worked—information technology: office automation, computers, email, radio mail, laptops, word processors, CADs, telephones, voice mail, videoconferencing, faxes, internet, intranets, websites, and so on. There is an infinite variety of gadgets and networks and servers that helps us store information, manipulate information, do arithmetic faster, retrieve information, transmit information, receive information, and examine information—in written form, spoken form, picture form, virtual reality form. Technology gives us faster, surer information when we want it and where we want it, in whatever form we want it. Understanding the information and using it wisely, of course, is then up to you and me. Technology does not do that for us. We're on our own in developing the wisdom and knowledge and understanding to make the information useful.

That's my mental map of what most people, especially people who work in offices, are thinking and meaning when they talk about technology. But I checked out the definition of *technology* in *The American College Dictionary*, and here it is:

1A. *The application of science, especially to industrial or commercial objectives.*

1B. *The entire body of methods and materials used to achieve such industrial or commercial objectives.*

  2. *The body of knowledge available to a civilization that is of use in fashioning implements, practicing manual arts and skills, and* extracting *[emphasis added] or collecting materials.*

So there's quite a lot there that we don't find if we just look in the office: technology that's not electronic, that is not about storing, manipulating, sending, receiving, and examining information. There's chemical technology; mechanical technology; electrical, civil, aeronautical, and space technologies; construction, metallurgical, textile, nuclear, agricultural, and automotive technologies; now biotechnology; and so forth.

I illustrated the point for my Boston audience with an example: I told them that I run a manufacturing company that produced and sold $802 million worth of carpets, textiles, chemicals, and architectural flooring in 1995—and would likely sell $1 billion worth in 1996—for commercial and institutional interiors. We, too, have offices chock-full of office technology: mainframes, PCs, networks—you name it. And people who are hoteling and teaming, working anywhere, anytime. Information technology makes it all possible, hooking us up around the world.

But we also operate factories that process raw materials into finished, manufactured products that, happily, many members of my Boston audience routinely use and specify for others to use, and our raw material suppliers operate factories. And when we first examined the entire supply chain comprehensively, we found that in 1995 the technologies of our factories and our suppliers, together, extracted from the Earth and processed 1.224 billion pounds of material so we could produce that $802 million worth of products. That's 1.224 billion pounds of materials from the Earth's stored natural capital. I asked for that calculation and when the answer came back, I was staggered. I don't know how the number struck them in Boston, or how it strikes you reading this, but it made me want to throw up.

Of the roughly 1.2 billion pounds, I learned that about 400 million pounds was relatively abundant inorganic materials, mostly mined from the Earth's lithosphere (its crust), and 800 million was petro-based, coming from either oil, coal, or natural gas. And here's the thing that gagged me the most: Roughly two-thirds of that 800 million pounds of irreplaceable, non-renewable, exhaustible, precious natural resource was *burned up*— two-thirds!—to produce the energy to convert the other one-third, along with the 400 million pounds of inorganic material, into products—those products that my Boston friends, and others like them around the world, had specified for others to use or used in their own offices, hospitals, schools, airports, and other facilities. That fossil fuel, with its complex, precious, organic molecular structure, is gone forever—changed into carbon dioxide and other substances, many toxic, that were produced in the burning of it. These, of course, were dumped into the atmosphere to accumulate, and to contribute to global warming, to melting polar ice caps, and someday in the not-too-distant future to flooding coastal plains, such as much of Florida and, in the longer term, maybe even the streets of Boston (and New York, London, New Orleans, and other coastal cities). Meanwhile, we breathe what we burn to make our products and our livings.

Don't get me wrong. I let that Boston audience know that I appreciated their business! And that my company was committed to producing the best possible products to meet their specifications as efficiently as possible. *But really, this cannot go on indefinitely, can it?* Does anyone rationally think it can? My company's technologies and those of every other company I know of anywhere, in their present forms, are plundering the Earth. This cannot go on and on and on.

However, is anyone accusing me? No! No one but me. I stand convicted by me, myself, alone, and not by anyone else, as a plunderer of the Earth. But no, not by our civilization's definition; by our civilization's definition, I am a captain of industry. In the eyes of many people, I'm a kind of modern-day hero, an entrepreneur who founded a company that provides over seven thousand people with jobs that support them, many of their spouses, and more than twelve thousand children—altogether some twenty-five thousand people. Those people depend on the factories that consumed those materials! Anyway, hasn't Interface paid fair market prices for every pound of material it has bought and processed? Doesn't the market govern?

Yes, but does the market's price cover the cost? Well, let's see. Who has paid for the military power that has been projected into the Middle East to protect the oil at its source? Why, you have, in your taxes. Thank you very much. And who is paying for the damage done by storms, tornadoes, and hurricanes that result from global warming? Why, you are, of course, in your insurance premiums. Thank you again. And who will pay for the losses in Florida and the cost of the flooded, abandoned streets of Boston, New York, New Orleans, and London someday in the distant future? Future generations, your progeny, that's who. (William "Bill" McDonough, dean of the School of Architecture at the University of Virginia and a leading proponent of green architectural design for many years, calls this inter-generational tyranny, the worst form of remote tyranny, a kind of taxation without representation across the generations, levied by us on those yet unborn.) And who pays for the diseases caused by the toxic emissions all around us? Guess! Do you see how the revered market system of the first industrial revolution allows companies like mine to shift those costs to others, to externalize those costs, even to future generations?

In other words, the market, in its pricing of exchange value without regard to cost or use value, is, at the very least, opportunistic and permissive, if not dishonest. It will allow the externalization of any cost that an unwary, uncaring, or gullible public will permit to be externalized. It's caveat emptor in a perverse kind of way. My God! Am I a thief, too?

Yes, by the definition that I believe will come into use during the *next industrial revolution*. (I didn't originate that term. Writer Paul Hawken has called for "the next industrial revolution," an idea that, as you can see, I have latched onto because I agree with him that the first one is just not working out very well, even though I am as great a beneficiary of it as almost anyone.)

To my mind, and I think many agree, Rachel Carson, with her landmark book *Silent Spring*, started the *next* industrial revolution in 1962, by beginning the process of revealing that the first industrial revolution was ethically and intellectually heading for bankruptcy.[1] Her exposure of the dangers of pesticides began to peel the onion to reveal the abuses of the modern industrial system.

So by my own definition, I am a plunderer of the Earth and a thief—today a *legal* thief. The perverse tax laws, by failing to correct the errant market to internalize those externalities such as the costs of global warming and pollution, are my accomplices in crime. I am part of the endemic process that is going on at a frightening, accelerating rate worldwide to rob our children and their children, and theirs, and theirs, of their futures.

There is not an industrial company on Earth, and—I feel pretty safe in saying—not a company or institution of any kind (not even, as I told my Boston audience, an interior design firm), that is sustainable, in the sense of meeting its current needs without, in some measure, depriving future generations of the means of meeting their own. When the Earth runs out of finite, exhaustible resources and ecosystems collapse, our descendants will be left holding the empty bag. Someday people like me may be put in jail. But maybe, just maybe, the changes that accompany the next industrial revolution can keep my kind out of jail. I hope so, most assuredly.

As maybe you can tell, I've seen the light on this—a little late, admittedly. But I have challenged the people of Interface to make our company the first industrial company in the whole world to attain environmental sustainability, and then to become restorative. To me, to be restorative means to put back more than we take; to do good to the Earth, not just no harm. The way to become restorative, we think, is first to become sustainable ourselves and then to help or influence others toward sustainability. Later, I'll show you our map for getting there: the Interface model for a sustainable enterprise.

Back to "Planning for Tomorrow," the theme of that Boston discussion. When we think of the technologies of the future, sustainability—an issue of absolute, overriding importance for humankind—will depend on and require what I believe are the *really and truly* vital technologies: the technologies of the next industrial revolution, whether developed by us, by our suppliers, or by others like us. I don't believe we can go back to preindustrial days; we must go on to a better industrial revolution than the last one, and get it right this time.

But what does that mean? I have read Lester Thurow's view that we are already in the *third* industrial revolution.[2] He holds that the first was steam-powered; the second, electricity-powered; making possible the third, which is the information revolution, ushering in the information age. Clearly, all three stages have emerged with vastly different characteristics, and it can be argued that each was revolutionary in scope.

However, I take the view that they all share some fundamental characteristics that lump them together with an overarching, common theme. They were and remain an unsustainable phase in civilization's development. For example, someone still has to manufacture your ten-pound laptop computer, that icon of the information age. On an "all-in" basis, counting everything processed and distilled into those ten pounds, it weighs as much as forty thousand pounds, and its manufacturers, going all the way back to the mines (for materials) and wellheads (for energy), created huge abuse to the Earth through extractive and polluting processes to make it. Not much has changed over the years except the sophistication of the finished product. So I refer to all three of those stages collectively as the first industrial revolution, and I am calling for the next *truly revolutionary* industrial revolution. This time, to get it right, we must be certain it attains sustainability. We may not, as a species, have another chance. Time is short, as we shall see when we get to a discussion of geologic time.

At Interface we have undertaken a quest, first to become sustainable and then to become restorative. And we know, broadly, what it means for us. It's daunting. It's a mountain to climb that's higher than Everest. It means creating and adopting the technologies of the future—kinder, gentler technologies that emulate nature. That's where I think we will find the model.

Someone has said, *A computer, now that's mundane, but a tree, that's technology!* A tree operates on solar energy and lifts water in ways that seem to defy the laws of physics. When we understand how a whole forest works, and apply its myriad symbiotic relationships analogously to the design of industrial systems, we'll be on the right track. That right track will lead us to technologies that will enable us, for example, to operate our factories on solar energy. A halfway house for us may be fuel cell or gas turbine technologies. But ultimately, I believe we have to learn to operate off current income the way a forest does and, for that matter, the way we do in our businesses and households: not off capital—stored natural capital—but off current energy income. Solar energy is current energy income, arriving daily at the speed of light and in inexhaustible abundance from that marvelous fusion reactor just eight minutes away.

Those technologies of the future will enable us to feed our factories with recycled raw materials—closed-loop, recycled raw materials that come from harvesting the billions of square yards of carpets and textiles that have already been made: nylon face pile recycled into new nylon yarn to be made into new carpet; backing material recycled into new backing material for new carpet; and, in our textile business, polyester fabrics recycled into polyester fiber, to be made into new fabrics, closing the loop. We'll be using those precious organic molecules over and over in cyclical fashion, rather than sending them to landfills, or incinerating them, or downcycling them into lower-value forms by the linear processes of the first industrial revolution. Linear must go; cyclical must replace it. Cyclical is nature's way.

In nature there is no waste; one organism's waste is another's food. For our industrial process, so dependent on petrochemical, human-made raw materials, this means "technical food" to be reincarnated by recycling into the product's next life cycle, and the next. Of course, the recycling operations will have to be driven by renewable energy, too. Otherwise we will consume more fossil fuel for the energy to recycle than we will save in virgin petrochemical raw materials by recycling in the first place. We want a gain, not a net loss.

But if we get it right during the next industrial revolution, we will never have to take another drop of oil from the Earth for our products or industrial processes. That epitomizes my vision for Interface.

Those technologies of the future will enable us to send zero waste and scrap to the landfill. We're already well down this track at Interface. We have become disciplined and focused in all our businesses on what is sometimes called the low-hanging fruit—the easier savings to realize. We named this effort QUEST, an acronym for Quality Utilizing Employees' Suggestions and Teamwork. In the first three and a half years of this effort, we reduced total waste in our worldwide business by 40 percent, which saved $67 million (hard dollars), and those savings are paying the bills for all the rest of this revolution in our company. We are on our way to saving $80 million or more *per year* when we reach our goals.

We're redesigning our products for greater resource efficiency, too. For example, we are producing carpets with lighter face weights (less pile) and *better* durability. It sounds paradoxical, but it's actually working, in a measurable way. We're making carpets with lower pile heights and higher densities, utilizing carpet face constructions that wear better in high traffic but use less materials—a tiny but important step in "dematerializing" business and industry, an intriguing aspect of the next industrial revolution. The embodied energy *not* used in the nylon *not* consumed is enough to power the entire factory making the redesigned products—twice!

Those technologies of the future will enable us to operate without emitting anything into the air or water that hurts the ecosystem. We're just beginning to understand how incredibly difficult this will be, because the materials coming into our factories from our suppliers are replete with substances that never should have been taken from Earth's crust in the first place—as we shall see. But just imagine factories with no outlet pipes for effluent and no smokestacks because they don't need them! Don't you like that? Paul Hawken was one of the first people I heard articulate this concept.

Those technologies of the future must enable us to get our people and products from Point A to Point B in resource-efficient fashion. In our company alone, at any hour of the day, we have more than a thousand people on the move, while trucks and ships (and sometimes planes) deliver our products all over the world. Part of the solution will be Rocky Mountain Institute physicist Amory Lovins's hypercar. When Amory's super-lightweight, super-aerodynamic hypercar is using solar energy for electrolysis of water to extract hydrogen to power its fuel cells and a flywheel, magnetically levitating at one hundred thousand rpm, in lieu of a battery, or using an ultracapacitor with nothing moving and nothing to wear out, to store energy, and recapturing the energy generated in braking the car and sending it back to power electric motors on each wheel without any drivetrain to waste energy, we'll be getting there with an important technology of the next industrial revolution.

To complement and reinforce these new technologies, we will continue to sensitize and engage all seven thousand of our people in a common purpose, right down to the factory floor and right out there face-to-face with our customers, to do the thousands and thousands of little things—the environmentally sensitive things, energy saved here, pollution avoided there—that collectively are just as important as the five big things, those technologies of the future: solar energy, closed-loop recycling, zero waste, harmless emissions, and resource-efficient transportation.

Finally, I believe we must redesign commerce in the next industrial revolution, and redesign our role as manufacturers and suppliers of products and services. Already we are acquiring or forming alliances with the dealers and contractors that install and maintain our products, requiring an investment of some $100 million in the United States between 1995 and 1998. With these moves downstream into distribution, we are preparing to provide cyclical, "cradle-to-cradle" service to our customers, to be involved with them beyond the life of our products, into the next product reincarnation, and the next. The distribution system will, through reverse logistics, become, as well, a collection and recycling system, keeping those precious molecules moving through successive product life cycles.

In our reinvented commercial system, carpet need not be bought or sold at all. Leasing carpet, rather than selling it, and being responsible for it cradle-to-cradle is the future and the better way. Toward this end we've created and offered to the market the Evergreen Lease, the first-ever perpetual lease for carpet. We sell the services of the carpet—color, design, texture, warmth, acoustics, comfort underfoot, cleanliness, and improved indoor air quality—but not the carpet itself. The customer pays by the month for these services. In this way we make carpet into what Michael Braungart, a German chemist, termed "a product of service;" what Paul Hawken described as "licensing" in *The Ecology of Commerce*;[3] and what the President's Council on Sustainable Development called "extended product responsibility." Walter Stahel, Swiss engineer and economist, was perhaps the first person to conceptualize such a notion.[4] (There's more about the Evergreen Lease in chapter 2.)

Environmental sustainability, redefined for our purposes as taking nothing from the Earth that is not renewable and doing no harm to the biosphere, is ambitious. It is a mountain to climb, but we've begun the climb. Each of the seven broad initiatives we've undertaken—the five areas of new technologies, sensitized people, and reinvented commerce—is a face of that mountain. Teams all through our company in manufacturing locations on four continents are working together on hundreds of projects and technologies that are taking us up those seven faces toward sustainability, toward the summit of that mountain that is higher than Everest. We know we are on just the lowest slopes, but we believe we have found the direction that leads upward.

We've embraced The Natural Step, the frame of reference conceived by Karl-Henrik Robèrt of Sweden to define the system conditions of ecological sustainability, as a compass to guide our people up the mountain.[5] In the thousands and thousands of little things, The Natural Step is helping provide what we have termed the sensitivity hookup among our people, our communities, our customers, and our suppliers. We want to sensitize all our constituencies to the Earth's needs and to what sustainability truly means to all of us. We want to engage all of them in the climb.

We started this whole effort in our company on two fronts. The first was focused on waste reduction. That's the revolution we call QUEST. It's our total quality management program, and more; the emphasis is broad. We define waste as any cost that goes into our product that does not produce value for our customers. Value, of course, embraces product quality and more—aesthetics, utility, durability, resource efficiency. Since in pursuit of maximum value any waste is bad, we're measuring progress against a zero-based waste goal. A revolutionary notion itself, our definition of waste includes not just off-quality and scrap (the traditional notion of waste); it also means anything else we don't do right the first time—a misdirected shipment, a mispriced invoice, a bad debt, and so forth. In QUEST there is no such thing as "standard" waste or "allowable" off-quality. QUEST is measured in hard dollars and, as I said, we've taken 40 percent, or $67 million, out of our costs in three and a half years, on our way to a rate of more than $40 million *per year* of waste reduction by the end of 1998, and that much or more again when we actually get to zero waste. One quick result: Scrap to the landfills from our factories is down over 60 percent since the beginning of QUEST in 1995—in some factories, 80 percent.*

We've also begun to realize that conceptually it might even be possible to take waste, by its current definition, *below* zero as measured against our 1994 benchmark. If we substitute one form of energy (solar) for

---

*The company's current achievements, twenty years later, are addressed in part 2.

another (fossil), or one form of material (recycled) for another (virgin), we are making systemic changes that create, in effect, negative waste when measured against the old norms. If successful, we will have replaced the old system, now obsolete and shown in comparison to have been wasteful all along, with the new, non-wasteful system. So to give this new meaning to everyday activities, we have further changed our definition of waste in one category and declared *all* energy that is derived from fossil fuels to be waste, waste to be eliminated systematically, first through efficiency improvement and, eventually, to be replaced by renewable energy. Even the irreducible minimum of energy needed to drive our processes is waste by this definition, as long as it comes from non-renewable sources. QUEST *is* a revolution in operational philosophy.

The second parallel effort we've called EcoSense. It's focused on those other four major technologies of the future, together with the thousands of little things and the redesign of commerce. Measurement is more difficult for EcoSense. We're dealing here with "God's currency," not dollars, guilders, or pounds sterling—the field called EcoMetrics, a term I coined. Here's an example of EcoMetrics: How do you evaluate the following hypothetical trade-off? One product consumes ten pounds (per unit) of petrochemically derived material, a non-renewable resource. Another, functionally and aesthetically identical to the first, consumes only six pounds, substituting four pounds of abundant, benign, inorganic material, but through the addition of a chlorinated paraffin. That chlorine could be the precursor of a deadly dioxin. How does one judge the true cost or value (which is it?) of that *chlorinated* paraffin—in God's currency? That's EcoMetrics, the search for God's currency. It's perplexing—a scale that weighs such diverse factors as toxic waste, dioxin potential, aquifer depletion, carbon dioxide emissions, habitat destruction, non-renewable resource depletion, and embodied energy. EcoMetrics: We need God's own yardstick, and wisdom, to help us measure where we are, let us know which direction we're headed, and to tell us when we reach sustainability. Dollars and cents alone won't tell us.

In February 1996 we brought these two revolutionary efforts, QUEST (the hard-dollar effort) and EcoSense (the "God's currency" effort, measured by EcoMetrics), together. We merged the two task forces into one and formed eighteen teams with representatives from all of our businesses worldwide, each team with an assigned scope of investigation. It was a wonderful marriage. It is integrating these closely related efforts and positively changing our corporate culture because it is making us think differently about who we are and what we do. As my associate Mike Bertolucci has said, "It is as if you enter every room through a different door from the usual one, so different is the perspective from which you view every opportunity." I call it "piercing the veil" and finding on the other side a whole new world of opportunity and challenges. Today there are more than four hundred projects, from persuading our landlord to install compact fluorescent lightbulbs in our corporate headquarters office to creating new, sustainable businesses within our company.

Other companies, different from ours, will have to pursue different technologies, different from ours. In the next industrial revolution, I believe they must if they expect to survive. In the twenty-first century, as the revolution gathers speed, I believe the winners will be the resource-efficient. At whose expense will they win? At the expense of the resource-inefficient. Technology at its best, emulating nature, will eliminate the inefficient adapters.

Meanwhile the argument goes on between technophiles and technophobes, one saying technology will save us, the other saying technology is the enemy. I believe the next industrial revolution will reconcile these opposing points of view, because there is another way to express the differences between the first industrial revolution and the next. The well-known environmental impact equation, popularized by biologists Paul and Anne Ehrlich in their writings, declares that:[6]

$$I = P \times A \times T$$

In this equation I is environmental impact (bigger is worse), P is population, A is affluence, and T is technology. An increase in P, A, or T results in a greater (worse) environmental impact. Technology *is* part of the problem: the technophobes' position. But that is the technology of the first industrial revolution—call it T_1. Now the equation reads:

$$I = P \times A \times T_1$$

What a dilemma! T_1 is not the answer. T_1 will not lead us out of the environmental mess, no matter how vigorously the technophiles assert it will. The more technology we have, the greater (worse) the impact.

Remember that "ten-pound" laptop computer and the extractive, abusive processes that produced it?

But just what are the characteristics of T_1, the technologies of the first industrial revolution? For the most part, they are extractive (written right into the dictionary definition of technology), linear (take-make-waste), fossil-fuel-driven, focused on labor productivity (more production per worker), abusive, and wasteful—the destructive, voracious, consuming technologies of the first industrial revolution. And they are unsustainable.

But what if the characteristics of T were changed? Call it T_2, now, the technologies of the next industrial revolution. Let's say they were *renewable*, rather than extractive; *cyclical* (cradle-to-cradle), rather than linear; *solar- or hydrogen-driven*, rather than fossil-fuel-driven; focused on *resource productivity*, rather than labor productivity; and *benign* in their effects on the biosphere, rather than abusive. And what if they *emulated nature*, where there is no waste?

Mightn't it then be possible to restate the environmental impact equation as:

$$I = \frac{P \times A}{T_2}$$

Wow! Then the technophiles, the technophobes, the industrialists, and the environmentalists could be aligned and allied in their efforts to reinvent industry and civilization. Move T from the numerator to the denominator and we change the world as we have known it. The mathematically minded see it immediately. Now the more technology the better (less impact). Furthermore, it begins to put the billion unemployed people of Earth to work—working on increasing resource productivity; using an abundant resource, labor, to conserve diminishing natural resources. Technology becomes the friend of labor, not its enemy. Technology becomes part of the solution rather than part of the problem.

What will drive technology from the numerator to the denominator? I believe getting the prices right is the biggest part of the answer; that means tax shifts and, perhaps, new financial instruments such as tradable emission credits, to make pollution cost the polluter—in effect, a carbon tax. In any event it means eliminating the perverse incentives and getting the incentives right for innovation, correcting and redressing the market's fundamental dishonesty in externalizing societal costs, and harnessing honest free-market forces. If we can get the incentives right, entrepreneurs everywhere will thank Rachel Carson for starting it all. There are new fortunes to be made in the next industrial revolution.

But what in turn will drive the creation of tax shifts and other politically derived financial instruments? It seems to me that those will ultimately be driven by a public with a high sense of ethics, morality, a deep-seated love of Earth, and a longing for harmony with nature. When the marketplace, the people, show their appreciation for these qualities and vote with their pocketbooks for the early adopters, the people will be leading; the "good guys" will be winning in the marketplace and the polling booth; the rest of the political and business leaders will have to follow. As a politician once said, *Show me a parade and I'll gladly get in front of it*. So will business and industry respond to the demands of this new marketplace, and Earth will gain a reprieve.

CHAPTER TWO

A Spear in the Chest and Subsequent Events

From my perspective—presumptuous, perhaps—this journey, this climb, has become an epic story. I've been told by Dr. Zink that all epic stories begin *in medias res*. So think of what I said in chapter 1 as the middle of things for my company, Interface, and me. What about the beginning? How did we get "here" from "there"? What was "there"? What is "here"?

Please indulge me as I switch to a personal vein. Someone has said that everybody has just one story to tell, her or his own story. Here is part of mine.

I was born and grew up in the small, west Georgia town of West Point, the third of three sons—"Baby Ray." I am also a product of the Great Depression, the era into which I was born, and World War II, as well as the postwar era that was, of course, one of enormous prosperity and economic opportunity in America. However, that was still over the horizon in the mid-1930s, and times were tough. My father, William Henry Anderson, was the oldest of seven children and had been made, as was often expected of the eldest son in those days (the early decades of the twentieth century), to sacrifice his own education and go to work to help his father support the family. After the eighth grade he quit school and went to work to enable his sisters eventually to go on to college. Even as a child, that struck me as a grotesque waste. Though I never heard him complain, I believe he knew it was a waste. That awareness shaped his determination not to let his sons waste their lives. He made sure that my brothers and I got an education. One brother, Bill Jr., became a medical doctor; the other, Wiley, a teacher. I became an industrialist.

My mother, Ruth McGinty Anderson, was also one of seven children, the middle child of more enlightened parents. Her eldest brother excelled in the pulpit, and other brothers excelled in business. She became a school-teacher, but (again as was customary in those days) was not allowed to teach school after she married. So she practiced her profession on her three sons. I responded well and loved school.

I grew up with a book in one hand and a ball in the other. Whatever ball was in season—football, basketball, baseball, and softball, in succession—occupied my days, and books occupied my evenings. It was the football that ultimately assured my opportunity to go to college. I earned a football scholarship to Georgia Tech, being good enough as a running back to capture the eye of legendary coach Bobby Dodd during his golden years at Georgia Tech.

However, it was my friendship with books, together with the study discipline imparted by my mother and my teachers in the public schools of West Point, that transformed that scholarship opportunity into an excellent education.

Ironically, it was an eighth-grade experience with football, in 1947, that was a defining moment for me—at the same age my father had been when his life was essentially defined for him by his father. I was big for my thirteen years, weighing in at 142 pounds, and the high school coach, Carlton Lewis, urged me to go out for football to develop my skills and to gain experience practicing against the really big boys, the high school varsity team. Very soon I found myself scrimmaging against those guys—the future state champion team—grist for the mill, so to speak. Playing defensive linebacker one day, I encountered the biggest, hardest-charging of the varsity running backs, one who went on to become an all-state and college-level player, at the sideline—my head against his knee in a thunderous collision. It hurt us both, his knee and my head. I could look up and see the knot rising on my forehead with each beat of my heart. That ended practice for both of us that day.

As far as I was concerned, it also ended football for me forever. The next day I did not go to practice, but Coach Lewis would have none of that. He left the practice field, found a telephone in a nearby house, and called my father at work in the post office. My father, justifiably glad to have obtained secure employment (dating from Depression days when secure jobs were hard to find) and having risen to assistant postmaster, never left his job for anything.

But he did that day, and he walked the streets of our small town (we didn't own a car) until he found me. He shamed me completely for quitting, with the tongue-lashing of my life, and the next day I was back at football practice, with sore forehead and chastened spirit, duty-bound not to waste my life with a bad eighth-grade choice. I never liked football again, but I played hard and successfully until a shoulder injury my sophomore year at Tech ended my playing career. I laid down my last ball and, with the hand thus freed, took up preparing myself for business, holding on to that book in the other.

I loved my mother and my teachers; I loved but resented my father for making me do something I hated. In my heart I felt that he, already financially burdened with two sons' college educations, saw my budding athletic ability as the way to shed the third such burden. I hated my high school coach, but over time grew to respect him (and perhaps love him) for making me better than I wanted to be, and—in a truly defining way— teaching me, as he taught everyone who came under his influence on the athletic field, to compete.

No other lesson in my entire life has been more valuable than that one. It is no accident that a hallmark of Interface today is the never-say- die, never-ever-give-up attitude of its sales force in pursuit of the next "heartbeat" for our company, the next order, and the goodwill of the loyal customer that provides that heartbeat, again and again.

It was competition at the academic level, beginning in the fourth grade, that further prepared me to excel at Georgia Tech. A new child joined our tiny class that year, and from her first day, Barbara Adams knew more than I did—always just a little bit more. Eight years of intense, head-to-head struggle later, Barbara nosed me out by 0.1 on a scale of 100.0 for valedic- torian honors in our graduation exercise. That competition on the book side of things was just as valuable a lesson as Carlton Lewis's unrelenting pressure on the athletic field. Today Barbara Adams Mowat is head of academic programs at the Folger Shakespeare Library in Washington, DC, senior editor of *Shakespeare Quarterly*, co-editor of *The New Folger Library Shakespeare*, and America's foremost authority on Shakespeare, one of an elite handful in the entire world. I like to think I had something to do with that—just as she influenced me, stemming from that struggle between two friends to be number one.

I worked hard at Georgia Tech, made Tau Beta Pi engineering honorary society, and graduated in 1956 with highest honors and a bachelor's degree in industrial engineering. I spent the next seventeen years climbing the corporate ladder and preparing myself by gaining broad business experience (mostly subconsciously) to take the plunge and become an entrepreneur. In 1973 I was thirty-eight years old and had a very good job with a major corporation. I left that job and company and cut the corporate umbilical cord to found a new company to produce, of all things, free-lay carpet tiles. The act of cutting that cord required one of the two hardest decisions of my life, a decision that had developed over the course of nearly seven years.

I had done well quickly in my climb up the corporate ladder. Four years after graduation from Georgia Tech, after unexciting jobs with two other companies, I found myself on a fast track upward, thanks to being picked out of a crowd of new blood to which I had been recruited as a part of the rejuvenation of Callaway Mills Company, a Georgia-based textile manufacturer. In 1959 Fuller Callaway Jr., chairman and CEO, began the process, though few of us recognized it, of preparing his family-foundation-owned company to be sold. Nine years later the sale occurred, but in the meantime it was "Camelot" for the management team that Mr. Callaway had handpicked to lead the process of pumping up the company for sale. But Camelot came to an abrupt end for me when I got passed over in favor of another executive for the job I most wanted: to head one of the three operating divisions of Callaway Mills. I was stung! To put it bluntly, being passed over chapped my ass.

That was in 1966, and my job as vice president, staff manager—responsible for all the non-financial staff functions of a company with $80 million in sales—no longer was enough to satisfy my ambitions. Wounded by the decision that disrupted my climb and over which I felt I had no control, I began the psychological journey toward entrepreneurship—doing my own thing and making my own decisions, especially those affecting my own career destiny. Two years later, on April 1, 1968 (that's right, April Fools' Day), Callaway Mills was acquired by a much larger textile company, Deering Milliken. All of my staff functions were quickly absorbed into their counterpart departments at Milliken, and soon I was reassigned to become director of development of Milliken's floor covering business. Within a year I fell in love with a new idea, which I saw for the first time in June 1969 in Kidderminster, England—carpet tiles, a new concept for covering an office floor.

I took a leading role in helping Milliken bring carpet tiles to the United States from Europe. It was a major development project, and by 1972 Milliken was the established leader in the United States in this emerging niche market.

After nearly seven years of searching, triggered by that 1966 disappointment, I had found *my* thing, too. Carpet tiles were so right! So smart!

For a year I corresponded with the Kidderminster carpet manufacturing company, Carpets International (CI), and finally persuaded them to join me in a venture to bring their patented carpet tile technology to the United States. As I write this, it is just about twenty-five years ago that Jim Carpenter, CI's vice chairman, woke me up on a January morning in 1973, by calling from England to tell me that CI was ready to go with the venture that we had been discussing and planning. Only there were some provisos. I must leave Milliken without legal impediment. I must retain legal counsel and public accountants satisfactory to CI. I must arrange a banking facility for the debt portion of our planned capitalization. And I must raise $500,000 of equity to go with the $750,000 that CI was prepared to invest in the proposed fifty–fifty Anglo-American venture.

That last one was a tall undertaking, much taller than I first thought. During the latter stages of our planning in late 1972, Smith Lanier and Robert Avary, both friends from my hometown, had been very supportive. With Robert's high level of interest and wealth to back that up, Smith and I thought that raising $500,000 would not be a big problem. So on January 19, 1973, I invested the first $50,000 out of my life savings (which, at the time, totaled about $65,000) to fund the initial capitalization of Compact Carpets, Inc., the predecessor of the predecessor of Interface.

We thought that my ability to satisfy another of CI's provisos—that I leave Milliken without any legal impediment—might be more difficult. After all, that depended on Milliken as much as on me.

On February 1, 1973, I tendered my two weeks' notice to Milliken, and told them generally what I planned to do: to create a new company from scratch, with CI as a partner, to compete with them in the emerging new product/market niche, free-lay carpet tiles for the office of the future. For the rest of that week and the next no one spoke to me, until Friday, February 12, my last day, when two of my fellow managers walked into my office, closed the door, and sat down. They told me that they did not believe I could legally do what I was planning to do, that I would unavoidably use proprietary information I had acquired at Milliken in any such new venture, and that such information was covered by a broad secrecy agreement I had entered into as a condition of employment with Milliken (following Milliken's acquisition of Callaway). Milliken had acquired different, competitive carpet tile technology, but the technologies were close enough that their concerns were legitimate.

More defiantly than I felt, I replied, "The hell you say!" And I invited them to bring their lawyer to Atlanta to meet with mine, Carl Gable, the next week. We set a date—Tuesday, February 16, 1973, in Carl's office.

Over that weekend, Robert Avary dropped out on me as an investor. He had by then committed to be my "angel" and to take any part of the $450,000 balance-to-raise for which I could not line up other investors. But he called me on Saturday from the psychiatric ward of Emory University Hospital to tell me that his wife had filed for divorce, that his financial holdings were "tied up," and that he had checked himself into the hospital. "Your need for money could not have come at a worse time" was his heart-stopping pronouncement.

What to do? What a dilemma! The Milliken meeting coming up, my bridges burned, and now this! What a stomach-knotting turn of events.

That's when Smith stepped up like the champion he was, and still is, and said, "Don't worry, we'll get the money." Those words will remain branded in my brain for as long as I will live. "Don't worry, we'll get the money": the classic link between entrepreneur and capitalist.

So Carl and I met with the Milliken people Tuesday. Carl challenged their position with, "What do you mean? Are you saying we can't compete?"

They quickly backed off. "No, we didn't mean that." And they tried very hard to entice me back with promises to forgive everything and treat me as if I had never left. It was very tempting. These were turbulent waters. Carl found me a private room when I asked for time to think about it one more time, to "review the bidding," so to speak. It was Smith's admonition, "Don't worry, we'll get the money," that was the cornerstone of my decision in that lonely room to move ahead. It was a pivotal moment. I returned to the meeting room, told the waiting group that I was going ahead, and asked Carl to find a way to satisfy Milliken's discomfort.

Joe Kyle, whom I had recruited to head up manufacturing for the prospective venture, and I went on to England. We spent the next three weeks finalizing our manufacturing plan with the help of CI technicians and getting a final decision from Peter Anderson (chairman of CI, and no relation to me) to go forward with the investment. It was a close call for Peter. By the end of the three weeks, Carl Gable had worked out an agreement with Milliken not to oppose the venture legally, provided I agreed not to use or disclose their most sensitive proprietary information, boiled down through long negotiations to their fire-retardant PVC formulations. That agreement, together with faith in the abilities of his technical people to work out such details and his own burning desire to get his carpet tiles into the American market—after *three* aborted negotiations with other prospective American partners, including Milliken—persuaded Peter Anderson to take the plunge. I believe Peter took it over the objections of his closest advisers, including Jim Carpenter, his number two man, who had been my first supporter at CI. Undoubtedly, it was a close and gutsy call for Peter; sadly, he did not live to see the astonishing success that flowed from that decision. Interface came oh so close to *not being* on so many occasions; it's a miracle that it *is*.

However, even with CI aboard, we still had the money issue to deal with—the $450,000 of American equity to be raised. Joe Kyle and I came home in early March, and Smith and I went to work on cultivating prospective investors. Smith and his family led with the first $200,000 commitment. Together with my initial $50,000, which was financing our current expenses, we were halfway to our objective. We found ourselves subject to an SEC rule that limited the number of potential investors we could even *approach* with our private placement offering memorandum to twenty-five. One by one, carefully screening and rationing our twenty-five contacts jealously, we got eighteen people to invest the remaining $250,000.

On April 6, 1973, the full commitment in hand, including a banking arrangement with the First National Bank of Boston (another CI proviso met), the CI money came in and Interface was capitalized under names long since abandoned: Carpets International-Georgia, Inc., and its sister company, Carpets International-Georgia (Sales), Inc.—CIGI and CIGI (Sales) for short. CIGI was owned 51/49, CI/Americans; and CIGI (Sales) was owned 51/49, Americans/CI. With the equity evenly divided between the two, we had, effectively, a 50/50 venture with a built-in tie-breaker. In a loggerhead disagreement, we could go our separate ways.

My American backers even agreed to go a step further and enter into a voting trust agreement, giving me, in effect, the vote of their 30 percent. Together with the 20 percent I owned, including sweat equity, that made me an equal partner with CI.

Eight years later CIGI and CIGI (Sales) would be merged into Interface Flooring Systems, Inc., as the American investors were able to buy 10 of CI's 50 percent to make the venture 60/40, Americans/CI. Five years later we would take over CI completely, get their holdings of our stock back, sell off CI's worldwide operations, and buy Guilford of Maine (a textile manufacturer) with the salvaged proceeds. That series of transactions, spanning three years, is a business school case study in its own right.

To tie up a loose end, in due course (with no help whatsoever from me and very little from CI), Joe Kyle worked out our own fire-retardant formulations, and we were able to produce competitive products that met all the fire codes. To this day I have never told another soul what I learned about PVC formulations from Milliken.

Peter Anderson, Smith Lanier, Carl Gable, and Joe Kyle are, without a doubt, on the very short list of people about whom I can say without fear of contradiction: Without them, there would be no Interface.

Still, even with all that support, the risk was so frightening, like stepping off a cliff in the dark and not knowing whether your foot would land on solid ground or thin air. Furthermore, throughout those tense months, beginning in late 1972, it had become increasingly clear that my wife, Sug, did not want me to do this thing I was so intent on doing. In fact, she had become adamantly opposed. We had two children, Mary Anne, sixteen, and Harriet, twelve. College was coming soon. Anyway, didn't I have a good job? Couldn't I just be content with that? Why, why, why?

Good questions. To this day I don't have good answers, other than that I was driven to do it, to seize the opportunity to do my own thing. In the midst of a heated argument with Sug one night, while the whole process was still unsettled, I stormed into an adjacent room and fell to the floor in abject anguish, wrestling with myself: To do it or not to do it? I got up from that floor and shouted aloud to no one but myself, "By God! I'm going to do it!" It was the hardest decision of my life up to then.

On April 6, 1973, when our equity capital was in hand and our banking facility was in place, Interface was born. Everything to that point, beginning with a gleam in my eye, had been conception and gestation. The birth was long, difficult, and very painful and could only be called complete when I had the initial capital in hand.

The decision that surpassed the earlier one as the hardest of my life came ten years later when I filed for divorce. The bitter seeds were sown in those arguments in 1972 and 1973 and in the vehement opposition I encountered from my wife at this turning point in my life. From that time on we grew apart.

Sug had every reason to object. Her protests and sense of insecurity were completely justified, but her opposition served only to spur me on. Not only did I never want to have a boss again controlling my career, but I could not let myself fail and face *I told you so* from her. My self-fulfillment didn't come without a price. The mother of my children paid a price, too, for my shot at success on my own terms. The children themselves, with lives and families of their own, responded with great maturity to the divorce and have accepted Pat, my wife of fourteen years now, more as close friend than stepmother.

Born in maelstrom, so close to not being born at all, Interface is precious to me.

More about this idea with which I fell in love in June 1969: Carpet tiles, modular carpet, came in eighteen-inch squares (in Europe, fifty-centimeter squares) that could be installed without adhesive to gain the appearance of broadloom carpet but have the easy, flexible functionality of modularity. At that time, carpet tiles were just beginning to be used in American office buildings, where the electrical wiring was in the floor, the furniture was designed for open-plan systems, and the office was becoming computerized. The new concept was known, even in those days, as "the office of the future." The office of the future needed carpet tiles for easy access to wiring in the floor and for their everyday practicality. The timing and the product concept were perfect. So right, so smart!

The new venture was an entrepreneur's dream, except for the tension at home. We began with the idea that carpet tiles were a better mousetrap and the time was right for them in America; this was followed by not only satisfying our initial provisions and birthing the company but also acquiring a site and a building, equipping a factory, securing raw materials in a time of extreme scarcity, developing and producing those first American-made products, building an organization of people—Joe Kyle in manufacturing, Don Russell in marketing, Don Lee in administration—launching a sales and marketing effort into the teeth of the worst recession since 1929, working like hell—and surviving! I found pioneering a new product in a new market to be the most frightening and stressful, yet exhilarating and highly rewarding experience imaginable. Our major competitor? Who else but Milliken! It was an in-your-face time. Compete, compete, compete! Beat Milliken, beat Milliken! Horatio Alger's stories were no better than this one.

Survive, we did, and we prospered beyond anyone's wildest dreams. Today that company is global. We produce in twenty-nine manufacturing sites, located in the United States, Canada, the United Kingdom, Holland, Australia, Thailand, and (our newest factory) China. We sell our products in more than 110 countries. Sales in 1995 exceeded $800 million; in 1996 they topped $1 billion for the first time; in 1998 they will likely exceed $1.3 billion. We make and sell about 40 percent of all the carpet tiles used on Earth, and enjoy the largest market share in nearly every one of those 110 countries. After numerous acquisitions over the years, we also produce commercial broadloom carpets, textiles, chemicals, and architectural products (specifically raised access floors) with some of the great brand names in the commercial interiors industry: Interface, Bentley, Guilford of Maine, Prince Street Technologies, C-Tec, Heuga (in Europe), Intek, Toltec, Stevens Linen, and Camborne (in Europe also). We distribute our carpet products through our owned and licensed distribution channel, Re:Source Americas. Most recently we acquired the Firth and Vebe operations in Europe from UK-based Readicut PLC. Our brands are not household words because our products are not used where people live; they are used where people work. We are the world's largest producer of contract commercial carpet. It is an "only in America" success story, another case study yet to be written for aspiring entrepreneurs.

But as successful as it seems, Interface is flawed. It took me a very long time to realize it, though. For the first twenty-one years of our company's existence, I, for one, never gave one thought to what we were taking from the Earth or doing to the Earth, except to be sure we were in compliance and keeping ourselves "clean" in a regulatory sense and obeying the law, and to be sure we always had access to enough raw materials, mostly petrochemically derived, to meet our needs. We had very little environmental awareness. Until August 1994.

True, before that we had been developing, for fully ten years, a program called Envirosense. Working through a consortium of companies, Envirosense had been focused on indoor air quality (IAQ) and alleviating sick building syndrome and building-related illness such as Legionnaires' disease. This effort had been based on some proprietary chemistry we had acquired in the field of antimicrobials, called Intersept (yes, with an *s*). Intersept is an additive that, if incorporated into plastic materials, renders the surface of those materials self-sanitizing. So with Intersept, materials such as carpets, paints, fabrics, air filters, and air duct liners, as well as cooling coil and drip pan coatings, can be made to be more hygienic. Better hygiene leads to better air quality by reducing bacterial and fungal growth, contributing to healthier indoor environments by tackling the microbial contamination piece of the very complex IAQ equation. We were accomplishing good things in the field of IAQ. That and compliance were *it* for us, in terms of environmental sensitivity.

But then in August 1994, some of the people in Interface Research Corporation, our research arm, in response to customers who were beginning to ask what we were doing for the environment—questions for which we did not have adequate answers—decided to organize a task force with representatives from all of our businesses around the world. Our research people wanted to review Interface's company-wide, worldwide environmental position and begin to frame a response to those customer questions we could not answer very well. One of these associates, Jim Hartzfeld, suggested that the new task force ask me to make the keynote remarks, to kick off the task force's first meeting and give the group an environmental vision. Well, frankly, I didn't have a vision except "obey the law, comply, comply, comply," and I was very reluctant to accept the invitation. The idea that, while in compliance, we might be hurting the environment simply hadn't occurred to me. Though I had heard Henry Kissinger advocate as early as 1992 that "sustainable development" should become the galvanizing cause for the West, replacing the Cold War in the new era of peace, I had no idea what he had meant. (I wonder today if *he* knew.) So I sweated for three weeks over what I would say to that group.

Then, through what seemed to be pure serendipity, somebody sent me a book: Paul Hawken's *The Ecology of Commerce*. I read it, and it changed my life. It hit me right between the eyes. It was an epiphany. I wasn't halfway through it before I had the vision I was looking for, not only for that speech but for my company, *and* a powerful sense of urgency to correct the mistakes of the first industrial revolution. Hawken's message was a spear in my chest that is still there.

Later I came to realize that it had touched me for another reason as well. At age sixty I was beginning to look ahead subconsciously to a day that would come soon enough when I would be looking back at the company I would be leaving behind. What would my creation, this third child of mine, be when it grew to maturity? I was looking, without realizing it, for *that* vision, too. A child prodigy in its youth, would it become a virtuoso? What would that mean? These were and are strategically important questions to me, personally, as well as to Interface, Inc.—in the highest sense of the word *strategic*. I'm talking about *ultimate purpose*. There is no more strategic issue than that.

In preparing to make that kickoff speech, I went beyond compliance in a heartbeat. I incorporated many examples from *The Ecology of Commerce* to explain what is happening to the ecosystem, using Hawken's description of the reindeer of St. Matthew Island to illustrate such basic concepts as *carrying capacity*, *overshoot*, and *collapse*, and as an arresting, frightening metaphor for the Earth:

> *A haunting and oft-cited case of . . . an overshoot took place on St. Matthew Island in the Bering Sea in 1944 when 29 reindeer were imported. Specialists had calculated that the island could support 13 to 18 reindeer per square mile, or a total population of between 1,600 and 2,300 animals. By 1957, the population was 1,350; but by 1963, with no natural controls or predators, the population had exploded to 6,000. The original calculations had been correct; this number vastly exceeded carrying capacity and was soon decimated by disease and starvation. Such a drastic overshoot, however, did not lead to restabilization at a lower level, with [just] the "extra" reindeer dying off. Instead, the entire habitat was so damaged by the overshoot that the number of reindeer fell drastically below the original carrying capacity, and by 1966 there were only 42 reindeer alive on St. Matthew Island. The difference between ruminants and ourselves is that the resources used by the reindeer were grasses, trees, and shrubs and they eventually return, whereas many of the resources we are exploiting will not.*

I cited Hawken's litany of abuse of the Earth that we are witnessing in our times:

- The depletion of the Ogallala Aquifer, that great underground body of fresh water in the American Midwest, and the implications of that—namely famine right in our own country. All over the world our aquifers are being dehydrated or, worse, polluted.

- The worldwide loss of twenty-five billion tons of topsoil every year, equivalent to all the wheat fields of Australia disappearing, and a hungry world population increasing by ninety million a year (now more like eighty million, but still . . .)
- The usurpation of a disproportionate share of net primary production, the usable product of photosynthesis, by the human species—one species among millions of species taking nearly half for itself—and pushing the ecosystem toward overshoot and collapse for thousands, maybe millions, of species.
- The result: an alarming increase in the rate of species extinction, now between one thousand and ten thousand times the average rate since the mass extinction of the dinosaurs sixty-five million years ago. As many as a quarter of all species and plants, animals, and microorganisms on Earth—millions of species—are likely to be lost within a few decades; as many as three-quarters face extinction in the twenty-first century. "The Death of Birth," Hawken called it. That phrase brought tears to my eyes when I first read it. It was the very point of the spear. The *death* of *birth*? Species lost, never ever again to be born. In no way can this bode well for our own species, because we are fouling our own nest, too.
- The cutting of vast areas of natural forests in Brazil, a critical lobe of the Earth's lungs, to clear land to raise soybeans to feed cows in Germany to produce surplus butter and cheese that piles up in warehouses, while a million displaced forest people live in squalor in the *favelas* (ghettos) of Rio de Janeiro. (I cried openly when I read that, and I was astonished and saddened still further to actually see favelas on a recent visit to Rio.)
- Illnesses from pesticide poisoning numbering in the millions each year, with uncounted deaths resulting.

In making that first speech, I borrowed Hawken's ideas shamelessly. And I agreed completely with his central thesis: that business and industry, together the largest, wealthiest, most powerful, most pervasive institution on Earth, and the one doing the most damage, must take the lead in directing Earth away from the route it is on toward the abyss of human-made collapse. I gave that task force a kickoff speech that, frankly, surprised me, stunned them, and then galvanized all of us into action. With and through them we are energizing our whole company to step up to our responsibility to lead. Unless somebody leads, nobody will. That's axiomatic. I asked, "Why not us?" Their answer has become a tidal wave of change in our company.

I offered the task force a vision: Interface, the first name in industrial ecology, worldwide, through substance, not words. I gave them a mission: to convert Interface into a restorative enterprise, first to reach sustainability, then to become restorative—putting back more than we ourselves take and doing good to Earth, not just no harm—by helping or influencing others to reach toward sustainability. And I suggested a strategy (you know this one, at least in part): Reduce, reuse, reclaim, recycle (later we added a very important one, *redesign*), adopt best practices, advance and share them. Develop sustainable technologies and invest in them when it makes economic sense. Challenge our suppliers to do the same. And I encouraged them to pick the year by which Interface would achieve sustainability. Two days later they told me their target year: 2000. I'll be sixty-six that year, and would love to see it happen by then. Truthfully, I think they were overly ambitious and that it will take much longer. The enthusiasm of the moment, coupled with a generous measure of naïveté as to the magnitude of the undertaking, led to that initial, optimistic goal. However, I come from long-lived people. The view from the top of that mountain that is higher than Everest will be beautiful beyond words! I hope to live to see it.

We gave this effort the name I've already mentioned, EcoSense. We are taking EcoSense throughout our company, hoping to involve everyone. It's not easy to get seven thousand associates to accept a role in a cause to do the right thing, and I doubt that every single one actually has. I cannot dictate what someone will believe in her or his heart, and that's where every individual decision lies. I just keep urging. For the first year, that urging yielded only barely perceptible effects outside the initial core group, but then the momentum began to gather. Our people, one by one, caught the vision. For the last three years, the progress has been phenomenal.*

EcoSense is basically our *internally* focused effort to do what's right. But it's not just the right thing to do; it's also the *smart* thing for a manufacturing company that is as dependent as we are on non-renewable resources (petroleum, coal, and natural gas) for its raw materials and its energy-intensive processes. Like carpet tiles in the beginning, EcoSense

---

* *Editor's note:* See the original Interface sustainability checklist in the appendix.

is so right, so smart. Only now, so much more is at stake—orders of magnitude more. Bill Young, who represents our outside accountants and has been another ally for twenty-five years, has observed that survival, an early preoccupation for us, has taken on new meaning, also orders of magnitude more important, as we try now to do our small part for the survival of the species.

———————

I made other speeches in the months that followed the first one, patterned after that kickoff address, all to Interface people, to begin to bring them aboard. My first outside public speech was to a group of Georgia Tech alumni and faculty; afterward, one of the professors in the audience, Dr. David Clifton, sent me a copy of Daniel Quinn's book *Ishmael*.[1] I read it, then read it again. I've read it six times now, and I've bought and given away some five hundred copies, always with the admonition, "Pass it on!" I'm here to tell you that Hawken and Quinn, together, will not only change your life but make you understand why it should change. They did both for me. If you haven't already done so, read *Ishmael* to understand why the world is in this environmental mess. Hawken will tell you *what* the mess is; Quinn, *why*. Those two books should be packaged and sold together, retitled *The What and the Why of the Biggest Mess of All Time*.

In July 1995 I met Paul Hawken. What a guy! We've become good friends. He told me that before writing *The Ecology of Commerce*, he read over two hundred books and a thousand papers, altogether more than twenty-five million words on the environment. He's distilled a lot into that monumental work. It's worth your time to read it if you haven't. Daniel Quinn, too, has become a good friend. His later book, *The Story of B*,[2] was also powerful, the philosophical sequel to *Ishmael*, written in an entirely different but very compelling way. Still another volume from Quinn is just out: *My Ishmael*.[3] I'm reading it now, and loving it. Hawken has another one coming, too, co-authored with Amory Lovins and Hunter Lovins (Amory's then wife and associate at Rocky Mountain Institute). It is called *Natural Capitalism*, and it will be an important work.

There was so much to learn that first year! I continued to read, going back to Rachel Carson's seminal *Silent Spring*, Vice President Gore's *Earth in the Balance*,[4] *Beyond the Limits* by Meadows, Meadows, and Randers (really scary!),[5] Lester Brown's *Vital Signs*[6] and *State of the World*,[7] Joe

Romm's *Lean and Clean Management*,[8] and others: David Brower, and more recently the work of Karl-Henrik Robèrt of Sweden and the account of how he initiated The Natural Step. Since we've adopted The Natural Step, it's obvious we think it's important. I'll tell more about it later.

I devoured Herman Daly and John Cobb's book, *For the Common Good*, trying to get a grip on the economics of sustainability.[9] They propose a system of economics that recognizes external costs, such as the societal costs of greenhouse gases, and seeks means for internalizing them by, for example, calculating the true cost of oil and getting the price right. Their book proposes to turn traditional economics on its head by suggesting that the nominal economic decision unit, the enlightened, self-interest-guided individual, be replaced by "persons *in community*" (emphasis added); that decision making be based on premises such as a full world with physical constraints being pushed to their limits, rather than an empty world with no constraints; and a world of finite resources rather than infinite resources. Earth's vital signs suggest that these are timely ideas.

During that first year of learning and growing, I also met and came to know and love John Picard. John is consultant to the Southern California Gas Company's Energy Resource Center (ERC) building project. It's a landmark green construction demonstration building, and the influences of Picard and Paul Hawken can be seen all through it. Hawken's book brought John and me together, and I worked with John to devise something he and the gas company really wanted in this building, the first-ever in the history of the world (to my knowledge) perpetual lease for carpet.

We called it the Evergreen Lease. In this lease Interface, as the manufacturer, made the carpet with state-of-the-art recycled content and also took responsibility for installing and maintaining it. Not only do we clean it regularly, but because it is free-lay carpet tiles, we selectively replace worn and damaged areas, one eighteen-inch square at a time. We implement a rolling, progressive, continuous face-lift by periodically, over the years, replacing modules and, most important, recycling the carpet tiles that come up. We continue to own the carpet. Title for the carpet tiles never passes to the user, the ERC; it stays with us, the manufacturer, along with the ultimate liability for the used-up, exhausted carpet tiles. The gas company pays by the month for color, texture, warmth, beauty, acoustics, comfort underfoot, cleanliness, and healthier indoor air (Intersept is built in)—the services carpet delivers—and avoids the landfill disposal liability altogether; that's our problem, and we intend to convert that liability into an asset through closed-loop recycling. We deliver the benefits and the services of carpet, but continue to own the means of delivery—theoretically for as long as the building stands.

Here's the thing: The economic viability of the Evergreen Lease for us *and its ultimate value to the Earth* depend on our closing the loop. That is, we must be able to recycle *used* face fiber into *new* face fiber to be made into new carpet tiles, and *used* carpet tile backing into *new* carpet tile backing. We have yet to learn to do either economically or energy-efficiently, much less to drive the process with the renewable energy that is necessary for sustainability. So you might say that we're cantilevered a bit and out on a limb of sorts. But we will get there. It's key to achieving sustainability, along with thousands of little things and the other big ones, those technologies of the future that put T (technology) in the denominator, and other new ways of doing business yet to be envisioned.

The Evergreen Lease is a manifestation of the future, not just for carpet but for a wide range of manufactured durable goods. It's one example of how commerce can be redesigned for the twenty-first century to use abundant labor to reduce dependence on diminishing virgin resources, and to increase efficiency in resource usage by forcing manufacturers to think responsibly cradle-to-cradle. We're grateful to the ERC and John Picard for driving this concept to a reality and letting Interface be a participant. I'll add this footnote: For the Evergreen Lease concept to become broadly successful, not only must we master closed-loop recycling, but financial institutions must get outside their comfort zones, too, and become third-party participants, providing financial intermediation for this strange concept they've never before encountered, a lease without a term for a product with indeterminate salvage value. Who can say what the value of those salvaged molecules will be in years to come? That will depend on the price of the fossil fuel precursors for virgin materials, which most likely will depend on revolutionary tax policies that are not yet on Congress's radar scope.

During these four years I've continued to read, even the other side's point of view. Early on a friend took issue with me and disputed Hawken, Lester Brown, and others, calling them "alarmists." We had a friendly debate going for a while. He sent me Bast, Hill, and Rue's book, *Eco-Sanity*.[10] It's the other view, that of the "foot draggers," you might say. It says good science doesn't support the views of the alarmists; that the world has a 650-year supply of petroleum, not 50; that the concern over the ozone layer is misplaced and unfounded; that acid rain is a disproven theory; that global warming is, too; that problems with automobiles, nuclear power, and oil spills are past problems that are nearly solved; that pesticides and toxic chemicals are manageable problems; and that deforestation and resource depletion are problems limited mainly to developing countries.

There's another book out there: *The True State of the Planet*, edited by Ronald Bailey, that conveys a similar "the sky is *not* falling" message to "Chicken Little" environmentalists, so to speak.[11] It forecasts a coming age of abundance, says we can wait awhile on global warming to get the computer models perfected, claims that famine is a thing of the past for most of the world's people, and so forth. These people quote historical data effectively, put great faith in human intelligence going forward, and write persuasively. They will test your resolve. They shook mine at first.

Honest people of goodwill and with good intentions can disagree. Just as we have technophiles and technophobes, all with sound reasons for their positions, I suppose good people can even interpret the same data differently and reach opposite conclusions, without having to be branded as foot draggers or alarmists. But how do we reconcile all of this? Where's the truth? I struggled for a year to find some answers to these questions. Had I put my company on a hopeless, misguided tangent?

My second outside public address on this subject, given to the US Green Building Council at Big Sky, Montana, in August 1995, was titled "The Journey from There to Here: The Eco-Odyssey of a CEO." Environmentally speaking, "there" was where I had been in August 1994, before that first task force meeting, pushing Intersept and IAQ through the Envirosense Consortium to make a buck, and staying in compliance on all the rest. "Here" was where I was one year later, speaking at Big Sky, with an awakened, sensitized conscience—realizing, for example, that compliance could mean "as bad as the law allows"—and with an awakened, sensitized company, hoping to do what's right, after wrestling for a year with what the truth was in all of this, and looking for a reconciling statement.

Well, I felt I had found that reconciling statement in my year of wrestling, or at least the beginning of one. Here it is: Our planet is billions of years old and has billions of years to go; creation goes on. David Brower, the eighty-five-year-old former executive director of the Sierra Club, often quotes son Ken Brower's observation that a living planet is a rare thing, perhaps the rarest thing in the universe. Also, thanks to very sophisticated modern technology that allows us to read the past, David has put *us*, our history as a species, our agricultural revolution, and our industrial revolution, into thoughtful perspective by compressing all of geologic time, from the initial formation of Earth 4.5 billion years ago right up to now, into the six days of biblical creation.

Using that compressed time scale (one day = 750,000,000 years), Earth is formed out of the solar nebula at midnight, the beginning of the first day, Monday. All day Monday is spent getting geologically organized. There is no life until Tuesday morning, about 8 AM. Amazingly, life, beginning with that first spontaneous cell somewhere in the primordial oceans, lifts itself by its own bootstraps, and survives! The prokaryote bacteria appear quickly, then proliferate, into mind-bending diversity, ever more complex. About Tuesday noon the blue-green algae already begin to create the oil deposits.

Millions upon millions of species come during the week, and millions go. What begins as a very toxic and hostile environment gradually is detoxified and sweetened as each species, through its metabolic processes, prepares the hostile environment for the next species, and the next, gradually sweetening the Earth's evolving biosphere and preparing the way for those that preceded us, and for us.

Thursday morning, just after midnight, photosynthesis—gradually building since Tuesday—gets going in high gear. Oxygen begins to accumulate in the atmosphere, and the protective ozone shield begins to develop. Soon after, in the wee hours of Thursday, more complex eukaryote cells, like those that will come to make up our own bodies, appear. Life begins then to really flourish and evolve into more diverse and more complex forms.

By Saturday morning—the sixth day, the last day of creation—there's enough oxygen in the atmosphere and sufficient ozone shield in the stratosphere that the amphibians can come onto land, and there's been enough chlorophyll manufactured for the forests and other land vegetation to begin to form coal deposits.

Around four o'clock Saturday afternoon, the giant reptiles appear. They hang around for quite a long time as a class of animals goes, until 9:55 PM, nearly six hours. (That would be *really* long for a species. None has lasted that long, and our species isn't likely to, either!) Just a few minutes after they are gone, a bit after 10 PM Saturday, the primates appear. (Incidentally, the Grand Canyon begins to take shape only about sixteen minutes before midnight.) *Australopithecus*, the first species on that branch off the main primate branch, the one that eventually leads to us, shows up in Africa at 11:53 PM, seven minutes ago. *Homo sapiens sapiens* arrives at 11:59:54: That is us, arriving on the scene just six seconds ago! The last six seconds at the end of a very long week, that's how long we've been here.

"Let the party begin!" David Brower said. But the party becomes a binge when, just a little over one second ago, 1.2 seconds in geologic time, we (that is, our forebears) throw off the habits of hunting and gathering and settle down to become farmers, and begin to change and sacrifice the environment to suit, and feed, our appetites. A third of a second to midnight: Buddha. A quarter of a second to midnight: Jesus of Nazareth. One-fortieth of a second ago, the industrial revolution ushers in the age of technology; the party picks up steam, so to speak, and kicks off the great carbon blowout that will characterize the first industrial revolution. An eightieth of a second ago, we discover oil and the carbon blowout accelerates. One two-hundredth of a second ago, we learn how to split the atom and the party gets very dangerous, indeed. I would show a time line for this week, but the last one-fortieth of a second would not be discernible. If the time line were *one mile* long, the industrial revolution would occupy the last 0.003 inch! One human lifetime, about 0.001 inch.

And now it's midnight, the beginning of the seventh day. The Union of Concerned Scientists, numbering some two thousand (including more than one hundred Nobel laureates), told us in 1992 that we had "*one to a few decades*" (emphasis added) to reverse course. In other words, the next two-hundredth of a second will be decisive; the time since we learned to split the atom (I remember 1939, so less than one lifetime), that short span of time projected not backward, but into the future, will be decisive. God can afford to rest on the seventh day, but I do not believe we can. I believe Earth needs a miracle.

To put it another way, the ten thousand years since the agricultural revolution began is, say, five hundred generations. Fifty years of oil is two and a half generations' worth; 650 years is thirty-two and a half generations' worth. Seems like a big difference on our scale of observation, but whether we're living in the last 0.5 percent of an epoch or the last 6 percent of an epoch doesn't really much matter. Time is short. In a blink of God's eyes, in an instant of geologic time, the whole epoch will be over. Whether the Earth will run out of oil in 50 years or 650 years may seem like a big contradiction in conclusions reached, but either, in geologic time, is the snap of a finger. Just that quickly, the reindeer of St. Matthew Island take on deep and personal significance if we care about future generations.

Our life span is so short that it's like being in only two or three frames of a movie that has been running a long time and has a long time yet to run. Our time on Earth is just so brief that we don't see enough of the movie, can't even see the next scene, much less where it's all headed. But our few frames can have an enormous effect on the outcome of the movie. Not to trivialize through analogy, but I remember hearing a NASA scientist say once, talking about Apollo 11, the first man-on-the-moon expedition, that 90 percent of the time the spacecraft was off course. The critically important mid-course corrections made it possible to reach the moon, and that determined the outcome. I stand firmly convinced that Earth—no, humanity—is off course and desperately needs a mid-course correction.

Dear reader—if I may be so bold, fellow plunderer—we've done a lot of damage in one-fortieth of a second, with our technology, since the beginning of the first industrial revolution and, *to be sure*, created enormous wealth, prosperity and economic growth in our part of the world, the developed world, but at what price to Earth? At what price, measured in God's currency? What is God's currency? I don't know exactly. At Interface we're trying to figure it out, but it surely is not dollars. Common sense tells us that neither the damage nor the economic growth—growing out of technology *as we practice it today*—can continue indefinitely; that the ways of the last one-fortieth of a second cannot go on and on. The first industrial revolution is unsustainable. We must not go on denying it. There is a limit to what our finite Earth can supply—and endure.

If it is true, as I was taught in college Economics 101, that all wealth ultimately comes from the Earth, then it must follow that wealth creation at the cumulative expense of a finite Earth is not a sustainable process. What was taught in economics, circa 1952, needs rethinking. Can it be true wealth if it is stolen from our descendants by the long arm of remote tyranny? If it is created by consuming Earth's capital reserves? Could we run our businesses or our households that way for very long, consuming our capital? This notion of wealth is more appropriately described as our children's inheritance, entrusted to us for safekeeping, being squandered for today's excesses.

Human intelligence, through design, together with human labor and energy *from the sun through photosynthesis* (up to now coming mostly by way of Earth's stored fossil fuels) add form, order, and purpose to raw natural resources; that is, they add value—all value—that we then use up. The whole process of wealth creation, as it has been pursued for the most part, depends ultimately on raw natural resources, whether from the forest, the field, the mine, the oil well, or the ocean. The process of creating usable value in them is always driven by the energy of the sun, either stored or direct, but so far mostly from stored fossil fuels.

So how much can we expect the Earth to yield from its finite, solar-derived, stored resources? It is self-evident that there's a limit. Therefore, we must learn to create wealth from more efficient use of resources and, eventually, from utilizing current solar income rather than living off the past until its stores are exhausted and denied to future generations. If we don't, we—that is, our progeny—will surely go to the poorhouse, another way of referring to ecological collapse. Human intelligence can take us a long way, but ultimately it will not be able to create something out of nothing. Dwindling resources eventually will dwindle to zero, given enough time, and there's a lot of that yet to be for the Earth, about five billion years or more, though surely much, much less for our species.

Exacerbating this whole issue is another lesson from Economics 101, which defines the basic economic problem as the gap between what each of us has and what each of us wants, a gap that drives all economic "progress," because it can never be closed. That is, no matter what we have, we want more. That's human nature. More precisely, that's human nature in our culture, the culture that Daniel Quinn calls the "taker" culture. Compound that with an ever-growing population, with each person wanting and striving for more, and it becomes clear that we have to find new, sustainable ways to satisfy needs and wants, other than by taking and taking from Earth's limited capacity to provide from its stored natural capital, and other than by dumping our poison into her limited sinks. What could be more obvious from the perspective of geologic time?

We just have to begin where we are, not where we wish we were—and sooner, not later, according to the scientists—to take those first steps in the long journey to sustainability, and begin to dismantle the destructive, voracious, consuming technologies of the first industrial revolution. We must start to replace them with the kinder, gentler technologies of the next industrial revolution, moving T from the numerator to the denominator and making technology part of the solution. We must begin to reinvent business, commerce, and probably this whole civilization (something Daniel Quinn is trying hard to think through), and find ways to create wealth (or perhaps redefine wealth), meet needs, satisfy wants, and raise standards of living for all without taking them out of the Earth's hide.

Perhaps that has already begun as a tiny gesture when we beam somebody up on a videoconference screen rather than get on an airplane to burn our part of the jet fuel that powers the plane, with us, from here to there. I say that's a start. It's one of the ten thousand little things we can do to conserve resources, avoid pollution, and honor nature. It's a beginning, but we really need to get busy identifying and doing the other 9,999. Conscience demands it. Or would we have our great-grandchildren curse us?

Another part of the reconciling statement lies in a paradox. A colleague and I were talking one day about the contradictory positions of the two polarized schools that I've hesitantly called the "alarmists" and the "foot draggers." Let's express those positions in terms of *perception*, *action*, and *outcome*. The alarmist perceives Earth to be in crisis, sees our actions as totally inadequate, and predicts the outcome to be collapse. On the other hand the foot dragger perceives things as not so bad, even getting better, sees our actions as good enough, maybe too good—meaning expensive and misguided—and sees the outcome as an abundant future for all.

Here's the paradox: The surest way to realize the alarmists' outcome, collapse, is to accept the foot draggers' view of where we are and what we need to do. On the other hand, the surest way to realize the foot draggers' outcome, abundance, is to believe the alarmists' view that we are in trouble and have to change. "Thesis and antithesis, reconciled through synthesis," to quote my friend, economist Mark Sagoff of the University of Maryland. He said Hegel would like that, furthering as it does the Hegelian view of the process of history.

When I first became sensitized to the environmental crisis in 1994, it seemed to me that there was a pretty heated dialogue going on between those parties at opposite ends of the spectrum, the people I have called alarmists and foot draggers to dramatize their diametrically opposed views.

On one hand the alarmists seemed to be saying, *It's hopeless. Humankind is doomed, and when our species goes, millions more will go, too.*

On the other hand, the foot draggers seemed to be saying, *Nonsense, things were never better. The rivers are cleaner, the air is cleaner. Living standards will continue to rise, the world population will stabilize. Everything is going to be okay.*

To try to reconcile these opposing viewpoints, I turned to the geologic time scale to understand what a relatively short time *Homo sapiens sapiens* has been around and what shameful damage the species has done, especially in the last "one-fortieth second" since the industrial revolution began.

However, as I now listen carefully to the opposing arguments, I hear something different emerging. I hear the foot draggers saying that the *historical* record just doesn't support the claims of progressive deterioration becoming catastrophic degradation; that rivers no longer catch fire; that the particulate count in our cities' air is dropping. (That may be true in many places, but it's not true everywhere, as anyone who's been to Bangkok, Sao Paulo, Mexico City, or Beijing knows.) They say that proven oil reserves are increasing; that food supplies per person are increasing; that the alarmists just don't put enough faith in human intelligence, ingenuity, and survival instincts.

Meanwhile, the alarmists are saying that we've just got to do something, we can't sit on our hands; that tax policy is wrong at the governmental level; that we tax good things, such as income and property, things we would like to encourage rather than discourage by taxing them. Instead we should be taxing bad things, such as pollution, waste, and carbon dioxide production, things we should be discouraging rather than giving a free ride or even subsidizing. We could change that with some applied intelligence and political will. We could harness current solar income with applied intelligence and stop consuming stored natural capital. We could close the recycling loop and eliminate waste and toxic emissions with applied intelligence. We could change the way we live, redefine wealth, come together as communities—with applied intelligence.

Now the different views don't seem so diametrically opposed. They seem to represent degrees of difference. Both views put the burden on applied human intelligence to find solutions. It's just a matter of whose sense of urgency you adopt. It's not so much a question of which direction we go, but how fast.

If we superimpose on today's dialogue the reconciling geologic time scale, we must be appalled by the damage done by our species in such a short time. Whether we take a hundred years to repair it, or a thousand, either is still a blink of God's eyes, but fix it we must. In that sliver of geologic time it could be over for us. We must apply human intelligence to the problem, whether it's creating those new technologies, rethinking taxation policy, or changing our lifestyle. Leave everything to market forces? For sure, market forces can help fix things, if we first apply human intelligence to redress the market's indifference to externalities, and get the prices right so they cover the true, all-in costs, and get the incentives right, too, and introduce those rewards that will stimulate the innovations that add to the denominator in the equation:

$$I = \frac{P \times A}{T_2}$$

In other words, let's be smart. Common sense tells us the sooner, the better; the sooner we begin, the more likely we are to avoid the abyss of St. Matthew Island.

CHAPTER THREE

Doing Well by Doing Good

I started in the middle of my story at that conference in Boston. Now you've got the beginning of my EcoOdyssey. It is becoming an eco-epic. But what about the ending? Where will our few frames in this movie take us, out of crisis or deeper into crisis?

Crisis is an interesting word. The Chinese symbol for crisis is a combination of two characters: 危 danger and 机 opportunity.

In terms of these two components, danger and opportunity, I want to outline how we as a company are thinking about this crisis.

First the *danger*: I believe that the Earth is damaged and hurting—badly. There is so much happening that it is just frightening.

Paul Hawken has laid out for us a litany of disasters that are happening all around us. To those, as if they weren't enough, we can add such cheery facts as:

- The decline of grain production in absolute terms worldwide; per capita, in decline since 1984! Grain feeds us directly and indirectly through livestock.
- The decline of fish catch per capita—worldwide. In decline since 1987, our own New England fishing industry experiencing collapse is an up-close, immediate example.
- Creeping death for the inland seas—choking on pollution; fish catch in the Caspian Sea is down by 99 percent! Is this a microcosm of the Earth's oceans in the centuries ahead? The decades?
- One billion people looking for work but not finding jobs. Another billion people already living in starvation conditions. Another billion on the fringe, hanging on by their fingernails. Half of the Earth's people in serious trouble. We cannot escape the consequences of such inequity. So many interrelated factors—population, food, jobs, industrial production, standards of living, toxic waste, pollution,

global warming, habitat loss, species extinction—all interrelated in ways that, with present trends, point toward collapse for the human species. We cringe from the ravages of the AIDS epidemic and we wonder, what's next?

- The loss of the rain forests—at the rate of a football field every two seconds—*now, now*. Even while the Brazilian government says it's not so, the satellite photos say it is so. Millions of acres of Sumatra and Borneo's rain forests, lost to fires, attest to the facts. An estimated half of all the Earth's species live in the rain forests, on just 7 percent of Earth's surface.

- Disappearing wetlands—wetlands that provide nutrients for the beginning of the food chain, and you know who's at the other end. Us! The wetlands also are the cradle of speciation. Even smelly swamps have a vital role.

- A nuclear waste cleanup that is off the scale in terms of both horror and cost, once estimated eventually to reach $300 billion. The unfolding Russian and Ukrainian situation could make this two times or three times larger. But nobody knows how to begin. Even if we learn to store the waste safely, what language shall we use for safety instructions for people twenty thousand years from now? No language on Earth is that old, none has survived that long—and plutonium is a five-hundred-thousand-year problem.

- Greenhouse gases, global warming, stratospheric ozone depletion (increasing UV radiation, increasing incidence of skin cancer), rising global temperature, melting polar ice caps. The largest iceberg ever known broke loose from the Antarctic ice shelf in December 1995 (their summer). As it drifts into warm waters and melts, to the extent it was once over land, not floating and already displacing ocean water, ocean levels rise an infinitesimal bit. The cumulative effect over time—devastating! Now the British Antarctic Survey tells us that the eight-thousand-square-mile Larsen Ice Shelf is critically unstable and may collapse. Temperatures around the ice cap are rising five times as fast as the global average, and grass has been seen growing along the edges of the icy continent.

- The scientific debate about global warming and climate change is over when twenty-six hundred atmospheric scientists from all over the world agree and only a handful hold out in skeptical disagreement.

The debate is now political. The science is compelling. Global warming is real! The political debate can go on, but sea levels will rise in the twenty-first century. How much? The most likely amount, about twenty inches, will be enough for a strong storm to put much of Florida underwater. It is already too late, probably, to save parts of Florida and other low-lying coastal plains. Some nine thousand square miles of the United States appear destined to be lost in just the next hundred years, absent massive dike construction. We would fight World War III before allowing a foreign aggressor to take nine thousand square miles of the United States! What *will* the twenty-second century bring, when atmospheric concentrations of carbon dioxide exceed two times pre-industrial-revolution levels, where today's computer models stop? What *will* a "four times" world be like? So far, there is nothing in the offing, *not even the Kyoto agreement* when ratified as treaty, to prevent it from happening. Action is needed now, the earlier the better. The longer we wait, the higher the concentrations will go into the zone of unknowable consequences.

- Toxic waste dumps, Superfund sites, that defy known cleanup technologies.
- Acid rain, killing whole forests in Central Europe and, along with industrial, agricultural, and municipal pollution, poisoning lakes everywhere.
- Atmospheric ozone drifting from our cities to rural areas, adding to the forests' distress and adversely affecting crop yields—a major factor in determining whether China will be able to feed itself while industrializing. A China that cannot feed itself will be everybody's problem.
- Especially apropos businesses like ours, non-renewable resources being gobbled up at obscene rates and, for the most part, burned for energy and converted into carbon dioxide to exacerbate the greenhouse effect.

With every life support system comprising the biosphere stressed and in decline, some say we're killing the Earth. No, I think the Earth will survive. Life, too, will survive, I think, even if it is beaten back and reduced to the last prokaryote cell to start all over again, as it started with that first cell, on "Tuesday morning" 3.85 billion years ago. Yes, I think life will survive, and that we probably will, too, for a time at least—but the loss of three-quarters or more of the species that share this planet with us, over the course of the next century, would constitute an unforgivable crime against our children and their children, limiting their life-sustaining options in ways that we cannot even begin to imagine.

Fraser firs once thrived in the higher elevations of the Great Smoky Mountains of east Tennessee. Today 90 percent of them are gone, their remnants standing like ghostly pylons against the green backdrop of the mountain slopes; the survivors, as were the vanquished, ravaged by the balsam woolly adelgid, an exotic species of insect accidentally introduced from Europe. The US National Park Service, explaining the tragedy of the firs, concludes with this chilling observation: *An exotic species eventually exterminates its host; then it dies, too.* We should study what ultimately happens to exotic species, because it surely looks as if we are one, wreaking havoc on every living thing in our path.

The Earth will be around for another five billion years or more, but meanwhile it is hurting. We are killing species by taking more than our share, fouling our own nest, and in the process diminishing the quality of life for our species and perhaps even dooming ourselves to extinction. That sounds harsh, but many thoughtful people believe this, the unthinkable, is true. St. Matthew Island *is* a metaphor for Earth. Perhaps, too, is the Fraser fir. It's only a matter of time.

That is the *danger*.

I'm often asked to define the business case for sustainability. How about, for starters: survival? Without sustainability, our descendants will watch society disintegrate and markets evaporate before their eyes. We cannot live without the life support systems of the biosphere any more than the other species can, and we continue to seriously over-stress those systems. The stress must stop for society, much less business, to thrive.

Now what about the *opportunity*?

First, you must understand that I am focused on the tiny corner where I live and work. If I cannot make a difference there, I surely cannot make a difference anywhere else. I have thought long and hard, and strategically, about how to make a difference through Interface, and I have read a lot about what has gone wrong and what seems necessary to make it right. Once I understood what Rachel Carson started, I felt morally obligated to help advance her legacy. Once you understand this crisis, no thinking person can stand idly by and do nothing. Denial is alluring, even seductive, but once you get past denial, you know you must do whatever you can. Conscience demands it; psychological liabilities begin to accrue.

But who will lead? The question demands an answer. Who can do something about all of this? Paul Hawken said that it must be business and industry; that they (that is, we) must take the lead. I look in the mirror, and I think he's right.

It's not the church, sadly. Though there are some encouraging signs that this could change, the church doesn't quite get it yet, even though in my church we sing:

> *This is my Father's world,*
> *And to my listening ears*
> *All nature sings and 'round me rings*
> *The music of the spheres.*

Too often, still, the church dogmatically helps perpetuate the myth Daniel Quinn identifies in *Ishmael*, the myth that the Earth was made for humans to conquer and rule. We have lived that myth, Ishmael reminds us, and sure enough there the Earth lies at our feet, bloody, broken, and conquered. Good people with awakened hearts could change this.

It's not government. The government never seems to lead, it always seems to follow, waiting for the people to create the parade, though it has a vital role to play with taxation policy: increasing taxes on "bad" things and relieving taxes on "good" things. What if our income taxes were reduced and our gasoline taxes were increased, but in an overall revenue-neutral way? How much better off we and the Earth would be! How much better still if the price of a barrel of oil, through taxation if necessary, reflected its true costs, including all of its externalized costs. Yet at this writing, the subject of tax shifts is not on Congress's radar scope, much less the legislative agenda. I saw a great bumper sticker the other day: IF THE PEOPLE WILL LEAD, THE LEADERS WILL FOLLOW. It's time for the people to step up and create that parade the politicians will run to get in front of.

It's not education, though education has a very important role to play in raising awareness and sensitivity in students, helping to draw the map to sustainability for all the disciplines, providing critical research to get the facts straight, and integrating the disciplines in a holistic vision of a new civilization. My school, Georgia Tech, is taking a leading role. I'll come back to what that is and how it came to be.

But if it's true that only business and industry can lead effectively and quickly, then how do you move the largest institution on Earth when it's actually millions and millions of separate entities? I tell my associates at Interface that if we (not just I, but we) get together as a company and take the lead, we can set an example for the entire industrial world, by first examining and understanding some basic things that Paul Hawken has pointed out for us:

- What we take from the Earth (those 1.2 billion pounds in 1995).
- What we make and what we do to the Earth in the making of it (products, stack gases, effluents).
- What we waste along the way (waste in all its forms).

What we take, what we make, what we waste—first to understand, then to do something about it, showing that it is good business to challenge and change all of these with Earth's benefit as the controlling criterion.

To expand the business case for sustainability, I believe that through EcoSense and QUEST we are pioneering a new business paradigm for success: doing well by doing good. There are two sides to that coin: (1) doing well (in a strict, moneymaking, business sense), and (2) doing good. To be clear, we want to do good because it's the *right* thing to do. We have to get our hearts right. But doing good may, at first blush, seem altruistic, softhearted, softheaded, even unbusinesslike. (Has Anderson gone 'round the bend? Well, yes, to see what's around there on the other side. That's part of my job. Having seen, I know I have never felt anything else that I have ever known in business to be, at once, so right and so smart. The closest thing was that first feeling when I fell in love with the idea of carpet tiles, a notable time when, once before, I went around the bend.) Who cares about that tree-hugging stuff? Well, I do, and many (maybe most, or all) of our people either do or are beginning to. And our customers—especially the architectural and interior design communities—do. They generally want to do the right thing, and want to do business with companies that are doing the right thing. And a growing number of end user and original equipment manufacturer (OEM) customers do, too.

I believe we can *also do well* by doing good. After all, we went into business in the first place to do well. But how do we do *well* by doing *good*?

In three ways, I believe:

First, as I've already suggested, by earning our customers' goodwill and, hopefully, their predisposition to trade with us, to help us in this hard, hard climb. But to earn that goodwill we have to avoid greenwash. Do you understand the term? Think of whitewash being used to cover a rotten fence. We must avoid that cynical, holier-than-thou, superficial cloak of green insincerity—so obviously self-serving, promoting products as "green" when they are not. We must be genuine. Our actions must speak louder than our words. Greenwash (pseudo-green) is, and should be, business suicide. Our customers should and will see right through it.

Second, through achieving resource efficiency. Amory Lovins, the brilliant physicist at the Rocky Mountain Institute who is working on that super-efficient hypercar, used the automobile to illustrate the vast inefficiency of our industrial system. He said that the objective of the conventional automobile, which weighs about 4,000 pounds, is usually to deliver a cargo, averaging about 165 pounds, from Point A to Point B, maybe to pick up a 1-pound loaf of bread. What is the efficiency of that automobile? Lovins said that, with internal inefficiencies (feel the wasted heat from the engine block and the brakes), only 15 to 20 percent of the fuel energy reaches the wheels as traction. Most of that moves the car's weight; only a small fraction, the driver's. The net result: About 1 percent of the fuel energy moves the driver. The hypercar will increase that by a factor of ten—in time, more. So what's the point of the traditional approach of spending millions to improve the internal combustion engine's efficiency by 1 or 2 percent? That's the wrong problem, and its solution will have a minor impact. The National Academy of Engineering agrees, and has concluded, using similar logic, that the overall thermodynamic efficiency of our American economy is (are you sitting down?) about 2.5 percent. Europe and Asia aren't much better. The Western economy is a waste machine, producing 97 percent waste!

When Interface takes petroleum or natural gas from the Earth and, with the help of our suppliers, converts it to carpet that gets used once and goes to a landfill ten to fifteen years later, I think of that as equivalent to about 2.5 percent efficiency in the use of those precious organic molecules. When we have increased that efficiency a minimum of ten times, we'll still be only 25 percent efficient, with lots more room to improve. We are looking for those who will help us to do it. If you are one of our suppliers, or hope to be, take heed.

Where are we in this quest for resource efficiency? The $1 billion of sales we recorded in 1996 consumed 19 percent less material per dollar of sales than we consumed in 1995, reflecting both increasing efficiency and our shift toward services, especially downstream distribution. This happened while we were realizing record profits, which was not an unconnected coincidence. Cumulative progress over three years is an increase of about 22.5 percent in resource efficiency; our share price has tripled.

I'll say it again: I believe that in the twenty-first century, the most resource-efficient companies will win! The sustainable will win big when oil's price finally reflects its cost and is $100, even $200 per barrel. Someday the market (and the economists) will wake up and the price *will* reflect the cost. That's the day for which we, as a company, are preparing.

At whose expense will the resource-efficient win? At the expense of the resource-*in*efficient. So I tell my people: We will win and the Earth will win! The best win-win I can think of. Business, like technology, can emulate nature and eliminate the inefficient adapters.

The third way we can do well by doing good is by setting an example that other businesses cannot ignore. The target group to influence, *other businesses*, are also our customers or potential customers. If we do well enough through creating goodwill and becoming resource-efficient, to the point that we are kicking tail in the marketplace, then that is the example other companies will see and want to emulate. Maybe they will become converts *and*, hopefully, customers, too. Then a positive feedback loop, the snowball effect, will take hold. The more good we do, the more well we will do. We can do the most good by doing the most well. By doing well, we will do more good through example. Then still others will see, and there will be that positive feedback loop—one of the few that is good for Earth.

I believe our customers will help us if they believe we are sincere. They will buy from us, even invest in our stock, and help us climb that mountain—even climb it with us. It is very important that we succeed in the climb, that the example be persuasively appealing to business and industry everywhere. Other companies will not look at us because they are benevolent or altruistic or philanthropic, nor because we are, but because we are undeniably succeeding in a different and attractive way. As our customers

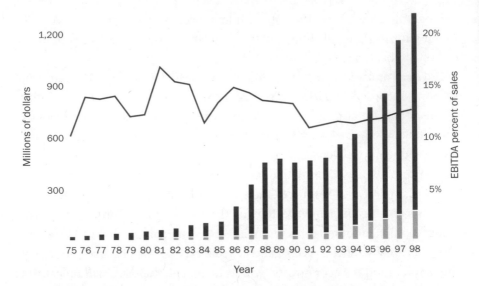

Interface, Inc., Annual Net Sales (*black bars*) and EBITDA (*grey bars*). EBITDA is earnings before interest, taxes, depreciation, and amortization (the measure of profits that is most important to banks and other lenders).

pitch in, our leverage with our suppliers will grow, as will our ability to bring them along on the climb. We cannot do it without both of them, customers and suppliers. Eventually, investors will see, too, that the only way for companies to do well in the long term is by doing the commensurate amount of good. Investors' capital speaks loudly, indeed!

That is the *opportunity*: to do well by doing good, and to make a difference by example—on a global scale—by making a difference in the corner where we live and work and inviting others to take a look and join in. Truly—to complete the business case for sustainability—entrepreneurs everywhere should thank Rachel Carson for starting it all; there are not only new but noble fortunes to be made in creating and bringing to market the technologies, the products, and the services of the next industrial revolution. Is it good business? Reach your own conclusion from our sales graph.

Of course, doing well by doing good and turning danger into opportunity are not the ending; they're only a tiny beginning of one possible ending. Who can say what the real ending will be? How long *Homo sapiens sapiens* will make it? Will going beyond the limits of growth that Dana

Meadows writes about in *Beyond the Limits* in fact soon lead to collapse and usher in an era of vast misery? Will self-declared wise, wise man be replaced in the "next few seconds" by a truly wise species?

It's safe to say, I believe, that if the developing world develops according to the model represented by the developed world, there will be unmitigated disaster in the twenty-first century. It helps sometimes to think in analogous terms. Most people in our industry understand what is meant by "footprint" of a building: the plan-view shape of the area a building occupies on its site. But we know that is not the entire footprint. There are also the mines from which the iron ore and other minerals came, the factories that produced the steel and glass, the transportation system that connects the building to its occupants, the power plant that supplies the building its electricity, the telephone company, and all the other supporting resources. When we think about it, they, too, are part of the building's footprint. Well, the footprint of a nation can be thought of as how much land it uses to meet all its needs. There are estimates that the United States already has a footprint larger than its geographic area. One indication of this is our balance of trade deficit. Let's say (though some think this very conservative) that the US footprint today is 1.2 times its geographic area; with 3 percent annual growth, we are on a course to double economically over the next twenty-five years.

At the same time, China, with a land area of about the same size as the contiguous forty-eight states of the United States, but with nearly five times the population, aspires to a standard of living for its people twenty-five years from now equivalent to ours today. What will its footprint then be? Well, something in the neighborhood of six times its land area. For example, if per capita automobile driving were the same in China as ours is today, China would consume the world's total gasoline production. So would India! And what about the rest of the developing world? Don't they, too, aspire to what the developed nations enjoy today—how can we deny them their aspirations? And the rest of the developed world, like the United States, continues to increase its take, also. Two comments from world leaders come to mind.

Maurice Strong, founder of the Earth Council, former chairman of Ontario Hydro, and special adviser to the secretary general of the United Nations, has said, "The fate of the Earth will be decided in the developing world." I think he is only partly right. I think Earth's fate is also in our hands because many of the solutions for the developing world must come from the developed world where the capital and expertise exist for developing those solutions. There are plenty of smart, creative people in the developing world, too, and they are watching our example as they decide how to deploy their more limited resources.

The second comment came from President Clinton. I was in the audience as he spoke at Georgetown University in October 1997, recounting how he had told China's president, Jiang Zemin, "The thing that I fear most is that China will get rich the same way we did in the West."

When the collective footprint of all the nations exceeds the area of the Earth, something will have to give. We must get either another planet or a new model. The current Western model that exploits the Earth (leaving T in the numerator) won't do, and we don't have much time to fix it. How do we begin? Taking that critical next step, in what direction shall we move? Toward what end? What shall we use as a map, as a compass?

CHAPTER FOUR

A Mountain to Climb

Deep into his book *Ishmael*, author Daniel Quinn, speaking through the teacher, Ishmael, used a metaphor to describe our civilization as it has arisen out of the first industrial revolution and the agricultural revolution before that. If you find yourself agreeing that maybe this civilization of ours is just not working out quite right, that something's seriously wrong, you will relate to Daniel Quinn's metaphor.

He likened our civilization to one of those early attempts to build the first airplane—the one with the flapping wings and the guy pedaling away to make the wings go. You've seen them in old film clips. In Quinn's metaphor, the man and the plane go off a very, very high cliff and the guy is pedaling away and the wings are flapping, the wind is in his face, and this poor fool thinks he's flying. But in fact, he is in free fall, and just doesn't realize it because the ground is so far away. Why is his plane not flying? Because it isn't built according to the laws of aerodynamics, and it is subject, like everything else, to the law of gravity.

Quinn said that our civilization is in free fall, too, for the same reason: It wasn't built according to the "laws of aerodynamics" for civilizations that would fly. We think we can just pedal harder and everything will be okay; pedal still harder and even fly to the stars to find salvation for the human race "out there." But we will surely crash instead, unless we redesign our craft—our civilization—according to the laws of aerodynamics for civilizations that would fly.

In the metaphor the very, very high cliff represents the seemingly unlimited resources we started with as a species and still had available to us when we threw off the habits of hunting and gathering, settled down to become farmers and, later, industrialists, and began to shape this civilization we have today. No wonder it took a while for the ground to come into sight.

But we are fortunate that there are people with better vision who have seen the ground rushing up toward us, perhaps sooner than most of us

have; and others who have undertaken to discover those laws of aerodynamics for civilizations that would fly. Wendell Berry, David Brower, Paul
Hawken, Daniel Quinn, Lester Brown, Dana Meadows, and others, carrying on the legacy of Rachel Carson, have seen the ground rushing up and
have sounded the alarm for all to hear. Buckminster Fuller, Walter Stahel,
Bill McDonough, Michael Braungart, Karl-Henrik Robèrt, and others have
set out to define the laws of aerodynamics for civilizations that would fly:
Fuller, recognizing that it's all a design problem; Stahel, McDonough, and
Braungart with "waste equals food" and "cradle-to-cradle" cyclical design
concepts; and Robèrt through his brilliantly conceived consensus document that laid out, for all to see, the science-based principles of sustainability that gave rise to The Natural Step. Let me tell you more about it.

Karl-Henrik Robèrt is a PhD and an MD whose field is research oncol-
ogy. He is highly respected in his native Sweden, especially for his work on
cancer in children. Kalle (that's his nickname, pronounced kŏl´e) said that
until about 1988, he had always thought that cancer was mostly a result of
lifestyle—in other words, people brought it on themselves through self-
indulgence. But when he saw an increasing incidence of cancer in children,
he realized that children didn't have such lifestyles, so he began to believe
that there were environmental causes. At the same time, he was struck by
the contradiction between the way individuals, acting as parents, would do
anything to help their stricken children, and the way those same people,
making up society and acting collectively, would do *nothing* proactively to
prevent the same tragedies. It seemed to Kalle as if group intelligence sank
below the level of the least intelligent individual in the group. He reasoned
that this was because the group was without a shared framework.

So Kalle set about to move an entire society to adopt a shared framework, to see the connections, and to do something about it. He decided
to try to create scientific consensus on the principles of sustainability to
provide that shared framework. It's a wonderful story, well told in Robèrt's
own publications, about how he came to achieve that consensus among a
peer group of more than fifty of Sweden's leading scientists, and then set
about to educate a whole nation about sustainability. Among the scientists
there was a great deal of disagreement, but in one area there was total, 100
percent consensus. This area of agreement was reduced to four fundamental principles. They have become recognized as the first order principles of
sustainability. You might call them the consensus-derived, science-based

laws of aerodynamics for civilizations that would fly—sustainable civilizations. Robèrt called these four principles the System Conditions for Sustainability. The "system" is the ecosystem, that thin shell of life where we and all the other creatures live, also called the ecosphere and the biosphere, which is eight thousand miles in diameter but only about ten miles thick— from sea level, five miles down into the depth of the ocean, and five miles up into the troposphere. Relative to a basketball-sized Earth, it is tissue-paper-thin and oh so fragile! For practical purposes, it sustains all life.

The principles of sustainability are based on scientific laws of nature that have been well understood for over a hundred years, the laws of thermodynamics. They are like the law of gravity. Someone has said, *They're not just a good idea, they are the law—the law of the universe.* Here are the first two laws of thermodynamics:

> **The first law** *of thermodynamics says that matter and energy cannot be created or destroyed. This is the principle of conservation of matter. When we burn something, it doesn't cease to exist. It changes form. When an automobile turns into a pile of rust, it doesn't cease to exist. It changes form. Every atom in the universe has always been in the universe. Every atom has existed since the beginning of time, and will exist until the end of time. It's true for matter; it's true for energy. Matter is energy. Neither can be created or destroyed.*
>
> **The second law** *of thermodynamics says that matter and energy tend to disperse. A drop of ink in a bathtub disperses. It may seem to disappear, but that's through dilution; it's still there, dispersed. Every manufactured article from the moment it takes its final form begins to disintegrate and disperse. A simple water glass, through the concentration of energy and design and human labor, is transformed from a pile of sand into a container, but from the moment of its completion it begins to disintegrate. If I drop it, I accelerate that disintegration. Another way to say it: the arrow of time flies in the direction of entropy, from order toward disorder. In a closed system, everything runs down. Everything that is concentrated eventually disperses.*

Matter and energy cannot be created or destroyed. Matter and energy tend to disperse. This means that any and all matter that is introduced into society will never cease to exist and will, sooner or later, find its way into our natural systems. It *will* find its way. It *will* disperse. Toxic materials are no exception. They, too, will disperse and find their way ultimately into our bodies. These are scientific principles. We can ignore them, but they will not go away. There are other laws of thermodynamics, but we can work with just these two for now.

Remember that process that started Tuesday morning (in geologic time)? Life lifting itself by its bootstraps, each species, through its metabolism, coupled with the sedimentation process, furthering the process of sweetening what started out as a toxic stew, preparing the way for the next species, and the next, and the next. It's important to understand that that toxic hostility has been, during the 3.85 billion years since "Tuesday morning," relegated to Earth's lithosphere—the crust. That's where it has been sent and sequestered through the inexorable process of sedimentation—down there; to make way for us, up here. Life has evolved in a self-reinforcing process: declining toxicity leading to more diversity, and more diversity leading to still less toxicity. New species after new species, a sweeter and sweeter Earth—and we are here today, the product of that self-reinforcing process.

What we have done in the one-fortieth of a second (in geologic time) of the industrial age is to reverse that process, to bring that stuff—lead, molybdenum, mercury, cadmium, antimony, copper, arsenic, asbestos, uranium, plutonium, hydrocarbons, and more—back to the surface in a zillion different forms, right into our living rooms, so to speak.

Though there were at least three earlier mass extinctions, a look at the "recent" history of Earth (the last hundred yards of our mile-long time line)—after amphibians came onto land and the evolutionary process was well along in the cleanup that prepared the way for us—reveals two cataclysmic reversals of the process. Something happened about 260 million years ago; scientists can only speculate about what environmental catastrophe created a "pollution spike." The spike was accompanied by a precipitous drop in biodiversity, a mass extinction, the most massive in Earth's history, as 96 percent of life vanished.

Then the forward evolutionary momentum was regained, and soon the giant reptiles evolved. Eventually they proliferated to rule the Earth until another pollution spike occurred, almost certainly this time from a comet or an asteroid that struck Earth 65 million years ago near present-day Yucatan and filled the atmosphere with toxic hostility. That time more than 75 percent of life became extinct, including the dinosaurs, but for their distant relatives, crocodiles and birds. The good news is, mammals got a chance.

Today, in the last 0.003 inch of that mile-long time line that represents Earth's entire existence, another spike has surely begun, and, with it, another mass extinction. Seventy percent of those polled in the five-thousand-member American Institute of Biological Sciences agree. This time, however, the cause cannot be attributed to unavoidable natural disaster but to the deliberate, willful, and quite unbelievable action of the highest form of evolutionary creature yet produced, and to the fruit of that species' intelligence—the industrial age. As before, the new spike, still in its infancy, shows increasing toxicity and is confirmed by the precipitous plummeting of biodiversity, as species disappear faster than anytime since the previous spike. Given the recent Asian oil discoveries, more efficient oil recovery technologies, and the extensive unmined deposits of coal, fossil fuels alone represent enough potential toxicity to kill us all if we foolishly continue to burn them for energy and dump the products of their combustion into the atmosphere. With all the other "stuff" coming daily from the lithosphere and being created by humans, the toxicity spike could go very high, indeed, and the loss of biodiversity could plummet a very long way.

Neither can we, as a species, adapt fast enough to tolerate such a drastic and sudden upheaval of the Earth's lithosphere. So children get cancer; older people get cancer, too, often from asbestos exposure or the like in their younger years. More subtle but nonetheless devastating things happen, too. Sperm counts decline, putting whole species, including ours, in long-term jeopardy. Likewise, pesticides (human-made, unnatural substances) kill people as well as insects; PCBs collect in fish and in human mothers' milk, both rendering it unsafe to ingest and putting endocrine systems at risk.

The self-reinforcing process can flip-flop and go the other way, too, and that has already begun, thanks to us, the human race. Increasing toxicity leads to loss of species. Loss of species (of trees, for example) leads to increasing toxicity, and still more species are lost. The "canaries in the mine" are warning us very clearly. We, as an endangered species, should be listening. Perhaps our endangerment stems from our unique ability among the species to engage in denial. Or is it simple arrogance that leads us to believe that it cannot happen to us? If we re-create the atmosphere of five hundred million years ago, we will not survive.

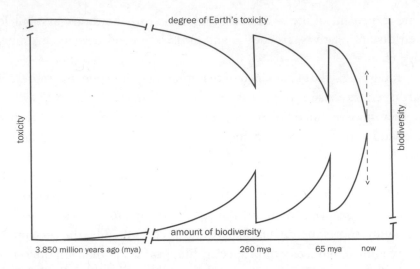

Self-Reinforcing Process. Conceptually, the mutually reinforcing relationship between toxicity and biodiversity is represented by mirror-image functions when the two variables are plotted against time. Though not to scale and disclaiming scientific precision, this graph depicts the global trouble in which we find ourselves and suggest there's more trouble ahead.

With that for background, here are the scientifically derived (based on the laws of thermodynamics) principles of sustainability, the system conditions of The Natural Step, published by Kalle and his consensus-reaching peer group, and ratified by American peer review as well. Please read carefully; this is what sustainability requires—absolutely, incontrovertibly. You cannot negotiate with a cell and tell it not to worry, that everything will be okay. Denial cuts no ice. It helps, too, to keep in mind the verity of geologic time. Here goes (and I quote extensively from the paper "A Compass for Sustainable Development" by Karl-Henrik Robèrt, Herman Daly, Paul Hawken, and John Holmberg):[1]

Substances from the Earth's crust must not systematically increase in the ecosphere. *This means that fossil fuels, metals, and other minerals must not be extracted at a faster pace than they can be redeposited and reintegrated into the Earth's crust, turned back into nature's building blocks. If substances from the Earth's crust systematically and inexorably accumulate, the concentration of those substances in the ecosphere will increase and eventually reach limits.*

We don't know what the limits are, but beyond the limits, irreversible changes will occur—so much radioactivity that we all die; so much lead in the water that we all become sterile; so much carbon dioxide in the atmosphere that the polar ice caps melt.

Substances produced by society (human-made materials) must not systematically increase in the ecosphere. *This means that human-made materials must not be produced at a faster pace than they can be broken down and integrated back into the cycles of nature, or deposited into the Earth's crust and turned back into nature's building blocks. If persistent human-made substances systematically and inexorably accumulate, the concentration of these substances in the ecosphere will increase and eventually reach limits. Again, we don't know the limits, but beyond those limits irreversible changes will occur. At some point, dioxins kill. Enough dioxins will kill us all. At some point, DDT, DDE, DES, mercury-containing compounds, and PCBs begin to disrupt endocrine systems. Endocrine systems keep our species going, and keep other species going as well.*

The productivity and diversity of nature must not be systematically diminished. *This means we cannot over-harvest or reduce our ecosystems in such a way that their productive capacity and diversity systematically diminish. We must certainly protect the small fraction of species that are capable of photosynthesis. We must not cut down the forests. They produce the oxygen that keeps us alive. Our health and prosperity depend on the capacity of nature to reconcentrate, restructure, and reorder building blocks into new resources. Rainforests and fisheries, farmlands and aquifers must not be pushed beyond their ability to recover. Species must be preserved; diversity in nature, protected. Why? Because we simply don't know all the interconnections in the web of life, but we know we are part of that web. It is foolish to say that we don't need this, or we don't need that. This and that lead to us.*

Therefore, in recognition of the first three conditions, there must be fair and efficient use of resources to meet human needs. *This means that basic human needs must be met in the most resource-efficient ways possible, and meeting basic needs for all must take precedence over providing luxuries for a few. Otherwise, we will reap a harvest of social as well as environmental instability. If people* *living in wooded or forested areas cut down all the trees for firewood because they don't have another source of fuel, all humanity suffers from the loss of biodiversity, and from the erosion, climate change, flooding, and desertification that follow. Fair is one thing, efficient is another; but they are intimately connected. How can we lift the lowest economically without dragging down the highest? The answer lies in resource efficiency.*

Resource efficiency is the rising tide that will float all the boats higher. If the National Academy of Engineering is anywhere near correct in estimating that the overall thermodynamic efficiency—in other words, the efficiency in the use of resources in the US economy to meet human needs—is 2.5 percent, then I say it again: Surely we must create for ourselves and show the developing world a better model than that. Technology (T) must move to the denominator, and we must remove *extracting* from the dictionary definition of technology! Technology must help put a billion unemployed people to work in gainful employment, conserving precious resources through cyclical, renewable processes.

The Natural Step is about reorganizing businesses and communities to conform to these Four System Conditions for Sustainability. These system conditions will define sustainability when we get there. But of course we are not there; we have a long way to go. How in the world do we get there when we, as a civilization, are headed in the other direction?

As we reach for these system conditions, our organizations must *systematically decrease* their economic dependence on underground metals and fuels and other minerals. Our organizations must *systematically decrease* their economic dependence on the production of persistent, unnatural, human-made substances. Our organizations must *systematically decrease* their economic dependence on activities that encroach on the productive parts of nature. And our organizations must *systematically decrease* their economic dependence on the use of unnecessary amounts of resources in relation to added human value; in other words, they must systematically move toward fair and efficient use of resources to meet all human needs, and put people to work to raise their standards of living, too.

This is The Natural Step. Karl-Henrik Robèrt created it in Sweden; Paul Hawken has brought it to the US; Jonathon Porritt has brought it to the U.K. Interface has publicly committed to be a Natural Step company. That means we have adopted these principles as our compass in our search for the path to sustainability, as our shared framework for understanding what sustainability is. Others (companies and individuals) are joining up, too. The Natural Step is becoming a force for good.

This is hard stuff. These are unrelenting principles, and they will not go away. Today we are violating every one of them in ways that must not go on. What about you? The laws of thermodynamics are undeniable, but no law says we must follow the principles of The Natural Step. That's a matter of choice. However, The Natural Step is telling us, in undeniable fashion, that we must, for the sake of humankind's future, replace T_1 with T_2 and move T to the denominator so:

$$I = \frac{P \times A}{T_2}$$

It is telling us further, in equally undeniable fashion, that social equity must be part of the fabric of sustainability.

Have I made the case for the next industrial revolution? For the urgent need to supplant industrialism, as we have grown up with it, with a better system? Why would I, along with so many others, say that the first industrial revolution was a mistake and is unsustainable? If that still seems completely outrageous, let me try a different tack.

Donella Meadows is one of the smartest people I know—an expert systems analyst, author, syndicated columnist, college professor, and farmer. Dana, as she is more casually known, has published an elegant paper, "Places to Intervene in a System."[2] In the style of a David Letterman Top 10 list, she listed these places to intervene in increasing order of effectiveness, beginning with number nine and working down to number one, the most effective place to intervene. Number nine on the list is to adjust the numbers: more of this, less of that. Working down the list, you find such things as adjusting the regulating negative feedback loops, driving positive feedback loops, and changing the goals of the systems. Number one on the list is challenging the mind-set behind the system—the paradigm, the perception of reality, the mental model of how things are that underlies the system in the first place. Dana said that this is the most effective place to intervene but also the hardest.

Now, we have systems all around us: Our transportation systems, our communication systems, our computer systems, our production planning systems, our systems of government, our accounting systems, our educational systems, our systems for managing our households, our regulatory system, our banking system, and . . . the industrial system that has arisen out of the industrial revolution.

What is the paradigm behind the modern industrial system? If you look at how it operates, you know it originated in another day and age, and it still views (or acts as if it views) reality as it did then:

- The Earth is an inexhaustible source of materials (natural resources). We'll never run out. There will always be substitutes available.
- The Earth is a limitless sink, able to assimilate our waste, no matter how poisonous, no matter how much.
- Relevant time frames are, maximum, the life of a human being; more likely, the working life of a human being; sometimes, especially in business, just the next quarter.
- The Earth was made for humans to conquer and rule; *Homo sapiens sapiens* (self-named "wise, wise man") doesn't really need the other species, except for food, fiber, and fuel, and maybe shade.
- Technology is omnipotent, especially when coupled with human intelligence, specifically left-brained intelligence (practical, objective, realistic, numbers-driven, results-oriented, unemotional); these will suffice, thank you very much.
- And (how about this one?) Adam Smith's "invisible hand" of the market is an honest broker.

Paul Hawken's *The Ecology of Commerce* and other books I've read since, together with my own late-blooming common sense, have convinced me that every element of that paradigm is wrong, dead wrong, and that survival of our species depends on a new industrial system developing quickly, based on a new paradigm—a new and more accurate view of reality, one that acknowledges:

- The Earth is finite (see it from space; that's all there is!), both as a source (what it can provide) and as a sink (what it can assimilate and endure).
- There will come an end to the substitutes that are possible. You cannot substitute water for food, air for water, food for warmth, energy for air, air for food. Some things are complementary.

- Relevant time frames are geologic in scale. We must, at least, think beyond ourselves and our brief, puny time on Earth—so brief—and think of our species, not just ourselves, over geologic time.
- Humans were made for Earth, not the other way around, and the diversity of nature is crucially important in keeping the whole web of life, including us, going sustainably over geologic time.
- Technology must fundamentally change if it is to become part of the solution instead of continuing to be the major part of the problem. T_1 must be replaced by T_2, and T must move from the numerator to the denominator.
- The right side of the brain, the caring, nurturing, artistic, subjective, sensitive, emotional side (in business, the "soft side"), is at least as important as the left side, perhaps a good bit more important since it represents the human spirit.
- The market is opportunistic, if not outright dishonest, in its willingness to externalize any cost that an unwary, uncaring public will allow it to externalize. It must constantly be redressed to keep it honest. Does the price of a pack of cigarettes reflect its cost? A barrel of oil?

Though the proximate (indeed, immediate) life-threatening problem we face as a species is the deterioration of the life support systems that make up the biosphere, there is a problem behind that problem: the industrial system that has arisen out of the mind-set, the flawed view of reality, that underlies the system. So it must follow that this flawed view of reality is the ultimate problem. Survival of our species, therefore, depends most of all on changed minds. (Sound familiar? That's Daniel Quinn's point, too.) An industrial system based on an erroneous view of reality has no foundation and will crash and take us with it, like a civilization that is ignorant of the laws of aerodynamics.

So at Interface we have chosen to intervene in the system. We have chosen Dana Meadows's most difficult and most effective place to try to make a difference. The reinvention of Interface reflects the new and more accurate view of reality, a new mind-set for a new industrial system. We are going about this reinvention ambitiously, aspiring to become the sustainable corporate model for the next industrial revolution.

Pioneering the next industrial revolution is a tall order. So is reinventing civilization. But someone must. What Rachel Carson started must continue. Either we, as a species, do it and find harmony with nature, or nature will whittle us down to size. She is in charge, just as she was at St. Matthew Island. The Earth's resources *are* finite; the Earth's capacity to endure abuse *is* limited. We must find a way to acknowledge and respect those limits.

The journey, with so far to go, goes on. In a global economy of $40 trillion ($40,000 billion), it's very presumptuous, isn't it, for a little company headquartered in Atlanta, which started from scratch in 1973 and took twenty-four years to reach $1 billion in sales, to think it can intervene in so large and complex a system as industrialism itself? What makes us think we can make a difference on a global scale and on an issue of such overwhelming magnitude? I really don't *know* that we can, but Paul Hawken, John Picard, Bill McDonough, Michael Braungart, David Brower, Daniel Quinn, Amory Lovins, Hunter Lovins, Bernadette Cozart, Dana Meadows, Herman Daly, Lester Brown, Rachel Carson, Kalle Robèrt, and others have convinced me that we should try. Unless somebody leads, nobody will.

At the very *least* we will give our people and our company a higher cause and a long-range reason for being. Abraham Maslow, describing the hierarchy of human needs, said a higher cause is important, and I agree.[3] After compensation to meet their needs, according to Maslow, people want the opportunity to develop and grow personally and professionally. When compensation is sufficient and growth opportunity is satisfied, people want to work for a company that makes a difference, that serves a higher cause. At Interface, we have learned in a very practical way that the quest for sustainability, for the welfare of our children's children, is a powerful, binding force. It is that higher purpose.

At the very *most* (let's dream a little), we'll start a trickle that will influence others—maybe you—to start their trickles; when those trickles come together into rivulets, and rivulets become streams, and streams, rivers, something good can happen.

As we used to sing in Sunday school when I was a child, "Brighten the corner where you are, brighten the corner where you are." Many years after learning that Sunday school song, I was exposed to the writings of the eighteenth-century philosopher Immanuel Kant and his somewhat more sophisticated corollary, "the Categorical Imperative."[4] What a great cause in which to invoke Kant and his eighteenth-century admonition! To paraphrase: "Before you do something, consider what the consequences would be if everybody did it." If we all succeed, individually, in doing good for Mother Earth in the corners where we live and work, in setting the example for others, and in governing our actions by the Categorical Imperative—*What if everybody did it?*—we will be helping Daniel Quinn in his mission to change six billion minds, creating a river of change and giving the Earth that mid-course correction. I believe it's not an option. It is humanity's only hope.

I have used another simile to describe sustainability as a mountain to be climbed. Let me expound. I have this mental picture of a mountain that is higher than Everest. It rises steeply out of a jungle that surrounds it. Most of us, people and companies, are lost and wandering around in that jungle, and don't know the mountain exists at all. Rather, we are preoccupied with the threatening, competitive "animals" all around us. A few have sensed the upward slope of the mountain's foothills under their feet. Still fewer have decided to follow the upward slope to see where it leads. And a very few are far enough along to have had a glimpse of the mountain through the leaves of the trees, to realize what looms ahead and above. Very few indeed have set their eyes and wills on the summit.

I am thankful that the people of Interface are in that small group. We have found the mountain's seven faces that must be scaled. Moreover, we have a compass, The Natural Step, to help us stay focused and on track to the summit, and we're developing EcoMetrics to measure our progress. We have some wonderful mapmakers—our team of advisers (Paul Hawken, Bill McDonough, John Picard, Daniel Quinn, David Brower, Jonathon Porritt, Kalle Robèrt, Walter Stahel, Bill Browning, Amory Lovins, Hunter Lovins, Dana Meadows, and Bernadette Cozart)—who are helping us find the fingerholds and the footholds to stay on the path to the summit.

What will the view from there, from the vantage point of sustainability, be like? I believe it will be wonderful beyond description, and I hope to see it before I die. I also hope others, still lost in the jungle or just becoming conscious of the upward slope, even beginning to explore it, will hear our cries of joy through the foliage and rush ahead to follow our path, someday soon to join us at the summit (or better yet beat us there). There's room there for everyone, and certainly anyone headed in that direction welcomes the companionship, as we seek to create the prototypical company of the twenty-first century, of the next industrial revolution.

"What's that?" you ask.

CHAPTER FIVE

The Prototypical Company
of the Twenty-First Century

What will it look like, the prototypical company of the twenty-first century that I want Interface to become—this model for the sustainable enterprise of the next industrial revolution?

Some time ago I was watching the movie *Mindwalk*, based on Fritjof Capra's book *The Turning Point*, about the interconnectedness of all things.[1] Of course, Capra was talking about interconnectedness at the subatomic particle level, and that was my first exposure to the idea. I've continued to read on that subject, but the movie started me thinking on another level about Interface's interconnections—the linkages between Interface and its constituencies: how some were good and others were bad, and how the good ought to be strengthened and the bad eliminated, and how still other linkages should be added. This led to a series of schematics to help all our people understand how we are approaching this monstrously difficult climb.

Here's the Interface corporate logo [*editor's note:* from 1998]. The circle represents the whole world. The *I* represents Interface within that global context, as well as, in a more subtle way, its focus on commercial *interiors*. Internally, we think and talk a lot about what belongs inside that "Circle I": What constitutes Interface?

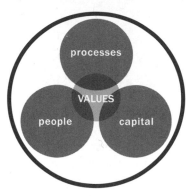

A typical version of what goes inside any company is people, capital, and processes. Economists often put *technology* where I have put *processes*. To my mind, *processes* is the broader word and the better choice. At the core are the company's values. The four elements vary specifically from company to company, but the general pattern holds for all. All companies, with their manifold distinctive differences, fit this general pattern. But of course, no company stands alone like this; each is connected to some important constituencies.

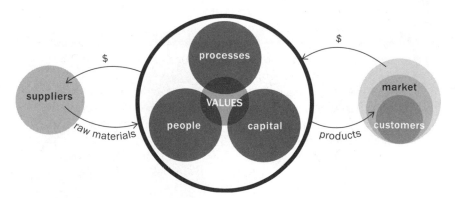

In our case Interface is part of a supply chain, with suppliers and customers and a market, our share of which we hope to increase. Products flow through that supply chain in one direction; money flows in the other direction. But the supply chain doesn't stand alone, either. It is connected to some other important constituencies.

Typical Company of the 20th Century. Our suppliers are dependent on the Earth's lithosphere for organic and inorganic materials. A very small amount of our raw material is natural, coming from the biosphere. Our processes are, unfortunately, connected to the Earth's biosphere by the waste streams and emissions we produce. And the products we make end up too often, at the ends of their useful lives, in landfills or, worse, in incinerators, creating a further pollution load for the Earth's biosphere to digest. Carpet in a landfill will last twenty thousand years.

We are connected to our community, too. Our people come from there, and their wages return to the community's economy; often they are its lifeblood. Our capital comes from a sector of the community, the financial sector; if we are fortunate enough to earn sufficient profits, dividends and capital appreciation are returned to those investors, along with interest to our lenders. Government is part of community, too. We are connected to it through laws, regulations, and, of course, the taxes we pay.

With these linkages in place, we have a description of many, many companies—in fact, almost every manufacturing company on Earth—and by analogy, of many other businesses and organizations. I have called this the Typical Company of the Twentieth Century. If this is all there is to Interface, Interface, too, is just typical.

However, we are trying to transform Interface into something different, a sustainable industrial enterprise. I call that enterprise the prototypical company of the twenty-first century. While the transformation goes on in every business throughout our company, Interface Research Corporation, the people who started the EcoSense effort within our company by convening that first task force meeting in August 1994 are leading us in the process of transforming our company. Let's see what that means, step by step. How do we get there from here and, in the process, pioneer the next industrial revolution?

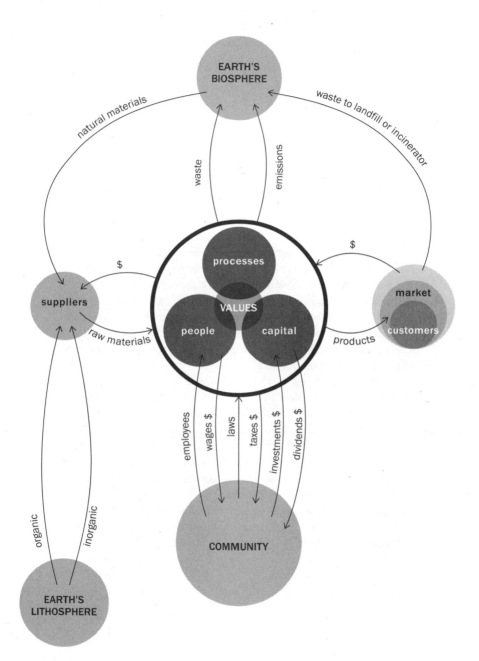

Front One: Zero Waste. We are pursuing the goal of creating the prototypical company of the twenty-first century simultaneously on seven fronts (the seven faces of our "mountain"). Though we are at different stages with each, we hope all will meet at the top. The first front is shown in this figure: Zero Waste. To attack unwanted linkages to the biosphere, we launched the effort we call QUEST, where any waste is bad, and anything we don't do right the first time is waste. Against ideal operational standards—zero waste—we identified $70 million in waste, based on 1994 operations. That's 10 percent of sales! We set out in 1995 on a mission to cut that in half by the end of 1997, then in half again by 2000, with hundreds of active projects, summed up and represented by the X in the diagram. We targeted $66 million of savings cumulatively over the first three-year period and much more in the ensuing years.

We did not get all the way there, but progress through fourteen quarters resulted in an index of 0.60, representing a 40 percent reduction in waste, saving cumulatively $67 million. This is real money, hard dollars, and it is paying for the rest of this revolution in our company. One quick result: Scrap to the landfills is down over 60 percent from 1994 levels throughout our company; 80 percent in some operations.

We have reframed QUEST for the next three years, through 2000, and we find, even after the savings that have already been achieved, that with a larger company there is now $80 million in waste. Once again, we intend to cut waste in half in three years, then in half again and again, until all waste is driven out of our company and the concept of waste is eliminated. That will require reinvention over and over again. We must be a learning company.

But when we get to zero waste, the savings will be much greater because our company will have continued to grow, and the opportunity will have grown. Further, as we redesign our products to use less and less material and to last longer and longer, we are dematerializing the business and reducing the load on the biosphere at the end of the supply chain.

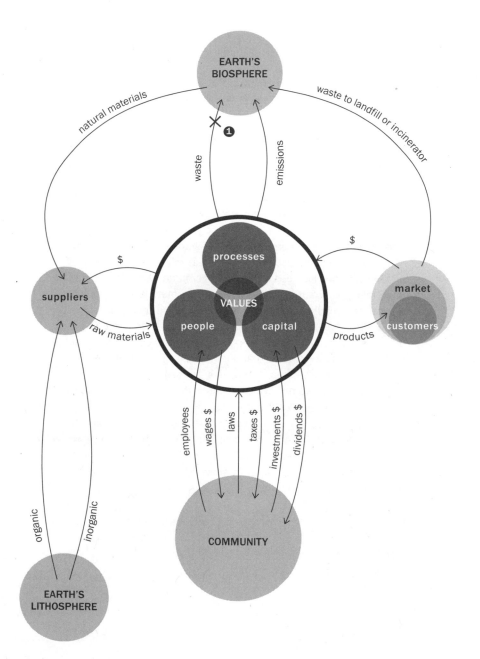

Front Two: Benign Emissions. The second front, Benign Emissions, attacks another unwanted linkage to the biosphere. We have inventoried every stack and every outlet pipe in our company, to see what is going out and how much of it there is, and we are reducing emissions daily. We have identified the world's most stringent regulatory standards and adopted them everywhere we operate. We began in 1995 with 192 stacks; with acquisitions since, that has become 229. Today we are at 184. There were eighteen process effluent pipes; now there are fifteen. In all, forty-eight stacks and pipes have been closed off. I hope to live to see the last stack and pipe closed off, in factories that don't need outlet pipes for their cyclical processes. Again, many projects are represented by each X.

But we know that to prevent *toxic* emissions altogether, we must go upstream and prevent toxic substances from entering our factories in the first place. What comes in will go out, one way or another. We are just beginning to understand how difficult that undertaking is. Commercially available raw materials are replete with substances that violate the first and second principles of The Natural Step (see chapter 4). Screening them out and remaining in business is a monumentally complicated undertaking. Yet we must. End-of-pipe solutions are unsustainable; they don't satisfy the principles of The Natural Step. Filters only concentrate the pollution, and then what do we do? We can't throw the filters away. There is no "away." Nothing is destroyed (first law of thermodynamics). It will disperse (second law of thermodynamics). Stopping pollution upstream is what we must do, leaving the toxic stuff in the lithosphere where the process of evolution, over the 3.85 billion years since "Tuesday morning," put it to make way for us. It must be left there (first system condition).

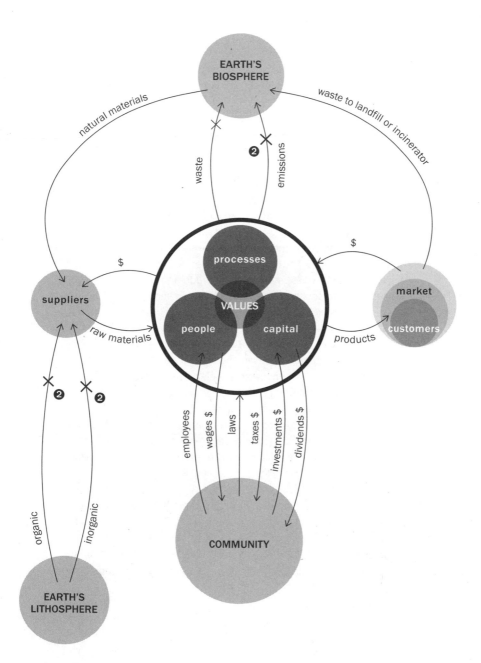

Front Three: Renewable Energy. The third front, Renewable Energy, means eventually harnessing solar energy. In the short term maybe hydrogen fuel cells or gas turbines, maybe wind (a form of solar), will run our processes, but eventually we believe it must be photovoltaic (pv) generated electricity. Harnessing renewable energy will attack numerous unwanted linkages, both to the lithosphere and to the biosphere, and will allow closed-loop recycling, the next front, to produce a net resource gain by obviating the need for fossil fuels for the energy to drive the recycling process. We have declared all fossil-fuel-derived energy to be waste and targeted it for elimination under QUEST—two fronts, hooked up! The initial emphasis is on efficiency. Amory Lovins's principles are our guidelines, and he is our mentor. Only when energy usage is at its irreducible minimum are we likely to be able to afford the investments in renewable sources. How far can we go? Farther than we ever imagined! In one case, through resizing pumps and pipes, we made a twelvefold reduction in connected horsepower for a key production line.

Our first application of photovoltaic power was in our Intek factory in Aberdeen, North Carolina. It is a 9 kWp (kilowatt peak) unit that runs one 10 hp motor at a cost of 32 cents per kWh. (The cost is primarily depreciation on the capital investment.) A better use of the pv power is to peak-shave electrical demand during the hottest part of the day when the air-conditioning load is greatest, and realize an effective cost of 15 cents per kWh, still four times the cost of fossil fuel electricity. Because it's not cost-effective, the pv array is a symbolic token. Greater savings are coming from natural daylight reflectors that track the sun from horizon to horizon to light the plant with daylight; the tracking is driven by a fraction of the pv generated power.

Yet we are pressing on. The next pv project is a 127 kWp unit in Southern California to produce the world's first solar-made carpet. We think solar-made will sell, that our specifier customers will love the idea. Who cares if the energy costs a little more, if the product sells and helps the Earth even a tiny bit? We're doing well and doing (a little bit of) good.

In Canada we have contracted with Ontario Hydro for green power (solar and wind). Even though it costs more, it's the right thing to do. We think solar-made carpet will sell in Canada, too.

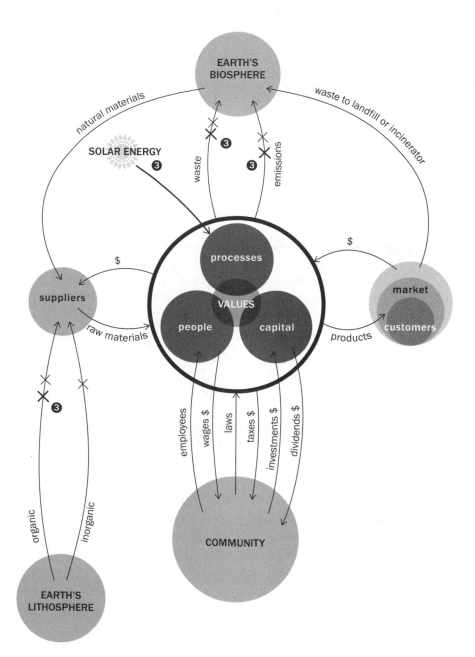

Front Four: Closing the Loop. The next front, Closing the Loop, introduces closed-loop recycling. Look at the impact this has on unwanted linkages! And see the new linkages that come into being. Two cycles are introduced: a natural, organic cycle, emphasizing natural raw materials and compostable products ("dust to dust"); and a technical cycle, giving human-made materials and precious organic molecules life after life after life through closed-loop recycling. The "sustainability link," the part of the technical cycle where closed-loop recycling will happen, must be invented and developed. It will be difficult and expensive to do, and we cannot do it alone. We need our suppliers' help here most of all.

But look at the power of it: The supply of recycled rather than virgin molecules in the technical loop, analogous to the supply of money in an economic system, will affect directly the resource-efficient "prosperity" of the enterprise. *What if everybody did it?* It would provide that rising tide that would lift the lowest on the economic scale, because recycling is labor-intensive. Labor for natural resources is a good trade-off that will get better as the prices (of petro-resources) get right.

This front goes hand in glove with the previous front, Renewable Energy. What's the gain if it takes more petro-stuff to create the process energy than is saved in virgin raw materials by recycling? Two more fronts, hooked up! If we can get both right, we'll never have to take another drop of oil from the Earth. That's the goal. It epitomizes our vision, along with factories with no outlet pipes—except that, unavoidably, the next front stands in the way.

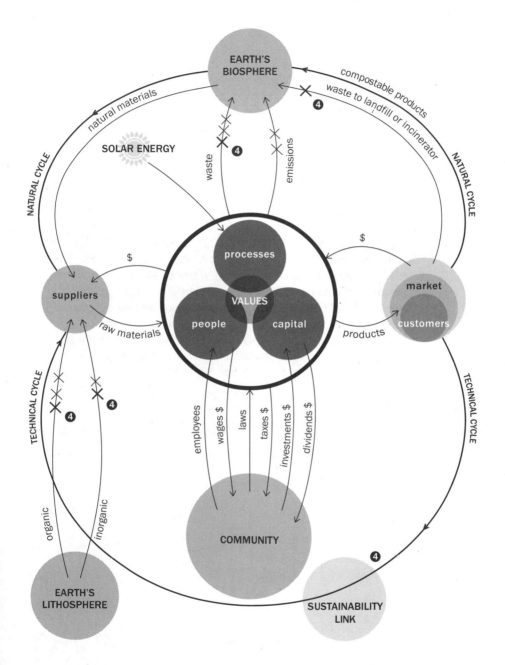

Front Five: Resource-Efficient Transportation. The fifth front is the one that is least within our control and the hardest for us to crack, especially with 100 percent sustainability as the ultimate goal. We can videoconference to avoid the unnecessary trip for a meeting, and we can drive the most efficient automobiles available. We can site our factories near the markets they serve, and plan logistics for maximum efficiency. But unless we choose to shut down contact with our customers and go out of business, we are dependent on the transportation industry for this one. Isn't everybody?

The good news is progress is being made—with electric cars, hybrid gas/electric cars, jet engines powered by hydrogen (coming from biomass or, someday, water), and hydrogen fuel cells that are advancing in efficiency and cost reduction. The "global brain" that physicist and author Peter Russell has described is waking up, and the transportation industry is part of it.[2] We need more "alarm clocks" to speed the process. Be one! We need Amory Lovins's hypercar. At the end of the day, we will have to resort to carbon offsets to completely resolve this one. We have already signed up with Trees for Travel, an organization planting trees in the rain forest, to close the gap. One tree over its life will sequester the carbon emitted in four thousand passenger miles of commercial air travel. We expect to plant a lot of trees.

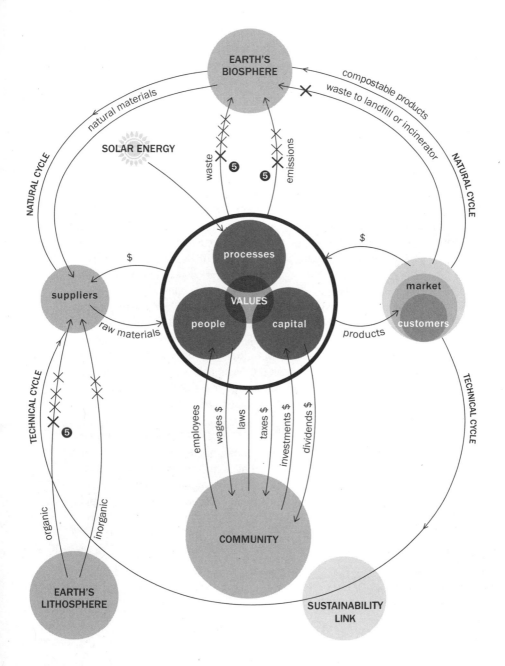

Front Six: The Sensitivity Hookup. Our sixth front, the Sensitivity Hookup, spawns numerous desirable connections: service to the community through involvement; investment in the community (especially in education); closer relations among ourselves (inside the circle) to get all of us in alignment, and with suppliers and customers. (I use *sensitivity* as Brian Swimme uses it in his book *The Universe Is a Green Dragon*, meaning heightened awareness brought about by absorbing a stimulus—an influence—and being changed in the process into a new person.)[3]

This front leads to increases among all, including our communities by way of our people, in the awareness of and sensitivity to the thousands and thousands of little things each of us can do to inch toward sustainability, breaking unwanted connections. Ties to the community, to our suppliers and customers, and within our organization are all strengthened. We hope our customers will see their role and become engaged in helping us increase our leverage with our suppliers to bring them along on the climb.

Community is redefined to include all of the community of life; our people are becoming sensitized to their stewardship responsibility for the treasure of life in all its forms, as well as the Earth's life support systems. So we've adopted streams and sponsored a television program to expose the plight of our own Chattahoochee River, one of Georgia's most polluted rivers. We're planting flower and vegetable gardens on our factory grounds and creating bird sanctuaries, too.

The Natural Step becomes at once our shared framework, our compass pointing the way, and a magnet, drawing us toward the summit of that mountain that is higher than Everest, called Sustainability. The ISO 14001 environmental management system is only a threshold—a given for all our factories. It will help us track our progress.

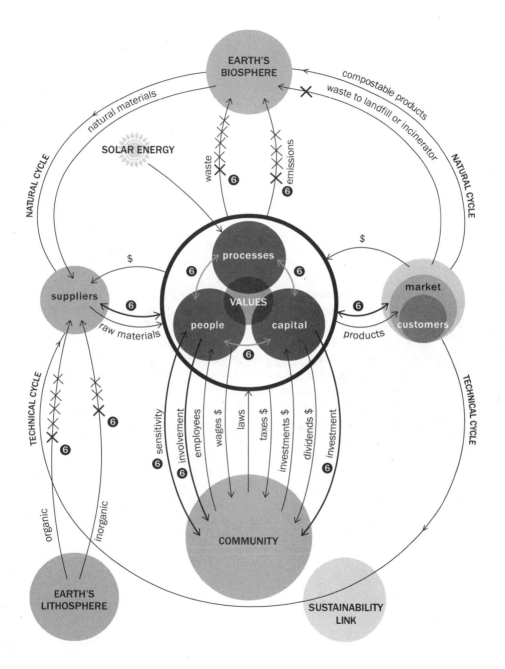

EARTH'S
BIOSPHERE

natural materials

compostable products
waste to landfill or incinerator

NATURAL CYCLE

NATURAL CYCLE

SOLAR ENERGY

waste

⑥

emissions

⑥

$

processes

⑥   ⑥

VALUES

suppliers

⑥

market

⑥

people   capital

⑥

customers

raw materials

products

$

TECHNICAL CYCLE

⑥ sensitivity

⑥ involvement

employees

wages $

laws

taxes $

investments $

dividends $

⑥ investment

TECHNICAL CYCLE

⑥

⑥

organic

inorganic

COMMUNITY

EARTH'S
LITHOSPHERE

SUSTAINABILITY
LINK

Front Seven: Redesign of Commerce. The seventh and final front calls for the redesign of commerce itself. This probably hinges, more than anything else, on the acceptance of entirely new notions of economics, especially prices that reflect full costs. To us, it means shifting emphasis from simply selling products to providing services; thus, our investment in downstream distribution, installation, maintenance, and recycling—all aimed at forming cradle-to-cradle relationships with customers and suppliers, relationships based on delivering, via the Evergreen Lease, the services our products provide, in lieu of the products themselves. As a result, we further break the undesirable linkages to the lithosphere and the biosphere, those that deplete or damage. Another highly desired result is increasing market share at the expense of inefficient competitors. But full-cost pricing is necessary if those salvaged molecules are to be, financially, worth salvaging to replace virgin petrochemicals.

Dream a little: Maybe even the tax laws eventually will shift taxes from good things, such as income and capital (things we want to encourage), to taxes from bad things like pollution, waste, and carbon dioxide emissions (things we want to discourage). What if perversity could once and for all be purged from the tax code? It must, for the next industrial revolution to put T (technology) in the denominator. When the price of oil reflects its true cost, we intend to be ready. That would truly change the world as we have known it, especially the world of commerce. It's also the day we'll be kicking tail in the marketplace.

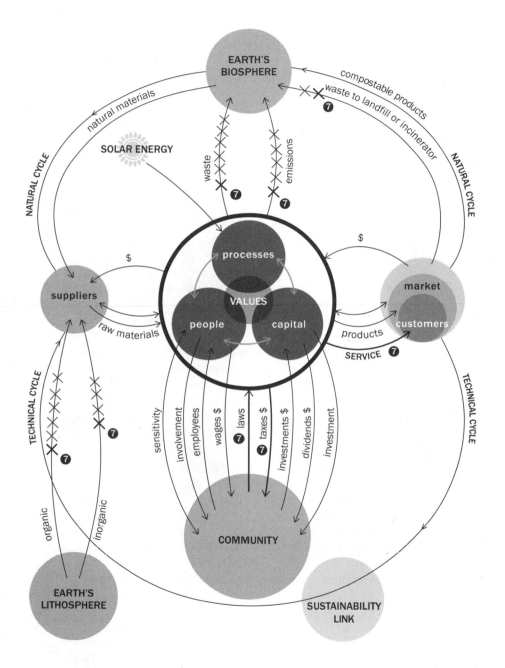

The Prototypical Company of the Twenty-First Century. Success on all seven fronts will bring us to our goal, the prototypical company of the twenty-first century. What are its characteristics? It is strongly service-oriented, resource-efficient, wasting nothing, solar-driven, cyclical (no longer take-make-waste linear), strongly connected to our constituencies—our communities (building social equity), our customers, and our suppliers—and to one another. Our communities are stronger and better educated, and the most qualified people are lining up to work for Interface. Customers prefer to deal with us, and suppliers embrace our vision.

Furthermore, this twenty-first-century company is way ahead of the regulatory process. The regulatory process has become irrelevant. The company's values have shifted, too, and it is successfully committed to taking nothing from the Earth's lithosphere that's not renewable, and doing no harm to her biosphere. The undesirable linkages are gone!

Sustainable and just, giving social equity its appropriate priority, and creating sustainable prosperity, an example for all, this company is doing well (very well) by doing good. And growing, too; it is expanding its market share at the expense of inefficient adapters, those competitors that remain committed to the old, outdated paradigm and dependent on the Earth's stored natural capital when oil's price finally reflects its cost ($100 per barrel, $200 per barrel, more?). The growth is occurring while extracted throughput (materials from the mine and wellhead) is declining, always declining, eventually to reach zero. Only zero extracted throughput is sustainable over geologic time.

It makes such absolute business sense to win this way, not at Earth's expense nor at our descendants' expense, but at the expense of inefficient competitors. Most important, we will have proven the feasibility of moving T (technology) from the numerator to the denominator, making technology part of the solution, and reducing environmental impact. If we can do that in a petro-intensive company such as Interface, anyone can do it. The next industrial revolution can happen.

In that new era the technophobes and the technophiles will be reconciled; the interest of labor and the interest of capital, reconciled; the interests of nature and the interests of business, reconciled. The Hegelian process of history—*thesis, antithesis, synthesis*—will lead to a sustainable society and a sustainable world. The mind-set behind the industrial system will have been transformed.

Will this company lose its competitive will to win? Not on your life! The prize is now larger than ever. We must never, ever give up—not on the next heartbeat, whether it is our company's or the planet's.

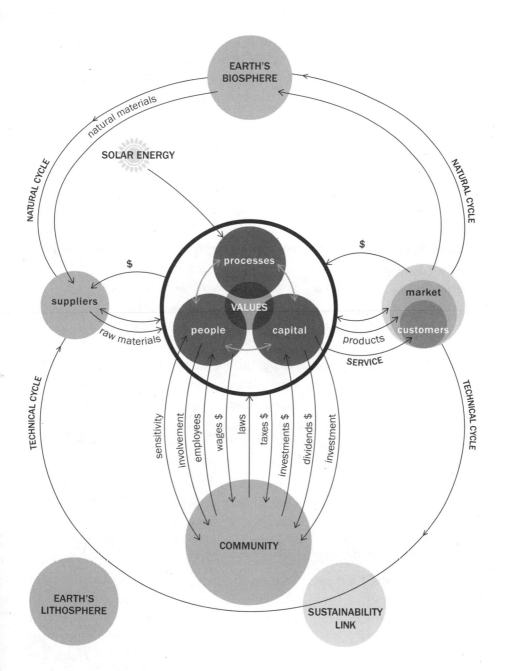

The tangible results of this seven-front assault, so far? I've already mentioned our supply chain's total extraction of materials from the Earth to produce 1995's sales of $802 million. That was 1.224 billion pounds of stuff from the Earth's crust, mostly petro-derived. On sales of $1 billion in 1996, our second year into this journey, the total extracted material was, by our very best calculation, very nearly the same at 1.23 billion pounds, or about 19 percent more efficient usage of the same amount of extracted material as in 1995. Furthermore, we're pretty sure that qualitatively what we produced in emissions and waste streams was at least no worse than the year before. All this means that our incremental sales volume of $200 million was achieved with practically no additional throughput of extracted material or harm to the biosphere, therefore at no additional cost to the Earth. Several factors contribute to this: a strong shift to service through downstream distribution, waste reduction through QUEST, product redesign to require less throughput, higher prices for our products (thus expanding profit margins), and various other product mix factors. Frankly, we were surprised by the apparent progress. Yet we believe the measurements are valid; this is the first $200 million or so of truly sustainable business that we have realized in this long climb, and it felt good!

Progress in 1997 was less dramatic, but did show further improvement in resource efficiency. With some estimating to establish our baseline year, 1994, after three years we find ourselves 22.5 percent of the way up that mountain. In three years the pounds of material extracted from the Earth and processed by our entire supply chain (including energy) has come down from 1.55 pounds per dollar of sales to 1.20 pounds per dollar. That's the 22.5 percent improvement—dematerialization—that, in 1997, yielded $256 million of sustainable sales. (See our results in the Progress Toward Sustainability and Total vs. Sustainable Sales graphs.) That felt good, too. It's a start, but the top of the mountain is a long way away. The top means zero extracted throughput per dollar of sales and no harm to the biosphere.

Meanwhile, the new thinking—the new mind-set—is beginning to permeate everything we do, especially product design and development. In our textile business, we've introduced fabrics produced from 100 percent recycled polyester and shifted entire product lines from virgin to recycled fiber. This was not accomplished without painstaking and excruciating effort on the part of ourselves and our suppliers. It's hard work. Further,

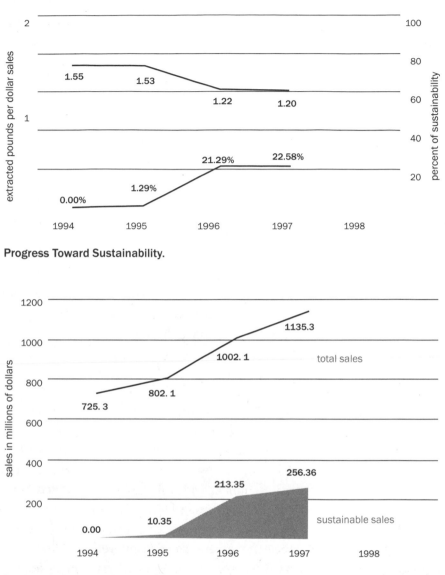

Progress Toward Sustainability.

Total vs. Sustainable Sales.

we've instigated the development of polyesters that use no antimony as a catalyst (leaving that poison in the lithosphere)—a significant technological breakthrough for our fiber producer/supplier. In our carpet tile business, a recent product introduction is produced by the fusion-bonded method (our initial carpet tile technology) with 72 percent closed-loop-recycled content—fiber-to-fiber, backing-to-backing. We call it Deja Vu, harking back to our beginnings twenty-five years ago.

Interface, Inc., made its first published progress report to the public in November 1997, the *Interface Sustainability Report*. It describes the problems of industrialism as we see them, the solutions we are pursuing, and where we are in our climb. It also makes clear how much farther we have to go. To our knowledge, it is the first such report ever produced by a corporation. It has been widely distributed to the company's stakeholders and is available to anyone else who requests one.

Other things need to change during the next industrial revolution. New technologies and manufacturing methods, tax shifts, and products of service are not enough. By every means possible, *extractive* through-put per unit of sales (gross domestic product on the national level, gross global product on the worldwide level) must be pushed toward zero. Sustainability depends ultimately on getting all the way there, to zero extractive throughput, given the perspective of geologic time and all the time yet to be. Again, we must remove the word *extracting* from the dictionary's definition of *technology*.

In October 1996 I was invited to Amsterdam to speak to the worldwide partners' meeting of one of the large international accounting firms. Since accountants tend to be highly analytical and unemotional, I thought that my standard speech, which has considerable emotional overtones, should be augmented if I wanted to connect with this group. So I added these comments just for the Amsterdam audience:

> *Now, what does this discussion have to do with you, your profession, and this meeting? I want to suggest that you and your profession are the scorekeepers in the game of business, but the rules of the game will change during the next industrial revolution; therefore, the method of scorekeeping will have to change, too, as business and commerce, and civilization, are reinvented. You could, with an early understanding of what might be, lead this change and help turn humankind from its course of self-destruction, unless of course you would rather just keep score as the world collapses around you. You could, for example, help to develop the field of EcoMetrics and*

help us understand God's currency, which certainly is not dollars or guilders, nor even pounds sterling.

I know that already you are faced with assessing environmental liabilities, but let's go further. For example, let us consider how we value assets today. Take a forest, a stand of trees. What is its value? I think most would say: x board feet of lumber at y $ per board foot equals z, less the cost of harvesting; that is the value.

But let me tell you a story about a small city on the banks of the Chattahoochee River in west-central Georgia in the United States, which in the first hundred years of its existence—through years of heavy rain and drought alike—never once experienced a flood. Then one year the banks of the river overflowed and $5 million of damage occurred. So the city fathers commissioned a dike to be built at a cost of $3 million, and that dike was sufficient to prevent flooding for five years. But then there was a season of especially hard rains, and the dike was breached, and the damage was $10 million this time. Therefore, the dike was rebuilt, higher this time, at a cost of $8 million, and the city was saved from flood for another seven years. And then, wouldn't you know it, the floods came higher still and the dike was breached again and someone finally said, What is going on here? So a team of experts was engaged to analyze the problem and one of the experts was an ecologist. And he, with brilliant insight, looked where? Not at rainfall records, nor at dike construction, nor at laminar or turbulent flows of a river. No, he looked upstream. And what did he find? He found that the forests for fifty miles upstream had been clear-cut over a period of twenty years and the clear-cutting had changed the hydrology of the area. Root systems no longer existed to hold the rainfall, so the rain ran off into the streams and rivers, eroding the land in the process and filling the river with silt and—by the way—killing fish, too, depriving the poor people of the area of one source of sustenance, while flooding the plains downstream, including the unfortunate small city.

So the question arises, What is the value of a forest? The short-sightedness of conventional economics lies exposed, naked, does it not? And I have not mentioned the value of a tree in removing carbon dioxide, a greenhouse gas, from the atmosphere, sequestering carbon, and producing oxygen for us to breathe, nor the songs of birds that are heard no more where the forests used to be. Neither have I mentioned the disease-spreading insects that now proliferate unchecked because the birds, their predators, are gone, resulting in an increase of encephalitis in the children in the region. So you see, there are serious questions to be raised about the traditional calculation of profit on the sale of the timber harvested from that clear-cut forest.

The ultimate solution to the flooding, pursued by our federal government in its dubious wisdom, was to build a dam at a cost of $100 million, which took twenty-eight thousand acres of prime agricultural land out of use and destroyed the habitat of uncounted creatures. Today the lake, thus created, is a polluted cesspool, collecting Atlanta's sewage. The value of a forest? Think again. (Though based on actual facts, the story is largely apocryphal and exaggerated, but I tell it for effect. I do know this river, though. As a boy I caught a twenty-pound channel catfish there that our family would eat for a week. Channel catfish no longer exist in the river.)

Or, staying with assets, what is the value of a mine, say a uranium mine—something that at first blush would seem to be highly treasured? But on second thought, when we consider the cost of the nuclear cleanup that Earth faces—somewhere between $300 billion and $900 billion, depending on just how bad the Russian and Ukrainian situations turn out to be—uranium somehow seems not to be so valuable anymore. Think of the liability we have transferred to future generations! Enlightened accounting would figure out how to take that liability right into the evaluation of that mining asset today.

Let's look at gross domestic product (GDP) for another exercise in new versus old economics. Consider, for example, that the Exxon Valdez *disaster in Prince William Sound added to GDP. Reflect on that. Reflect also on the absurdity of the fact that the medical expenses for a child dying of environmentally related cancer add to GDP. And that the costs to clean up and rebuild after a hurricane caused by global warming add to GDP. Clearly, as a measure of standard of living, much less as a measure of progress or well-being, GDP is sorely lacking.*

I spoke earlier about the cost of a barrel of oil, compared with its price, and how the market is oblivious to the notion of external costs, both those passed on to our neighbors and those passed on to our grandchildren, what I've called intergenerational tyranny. We must think more about present value discount rates. Perhaps they should be negative, *increasing the present value of future liabilities, rather than decreasing them. In taxation policy, the Earth cries out for a carbon tax to increase the price of fossil fuels to internalize the societal costs of military power in the Middle East and global warming, and thus hasten the development of alternative energy sources.*

Herman Daly, an economist at the University of Maryland in the United States, has been considered a kook by mainstream economists for years. Daly criticizes conventional economics as "empty world" economics and the economics of "unlimited resources" in what's clearly an emerging era of a "full world" with physical constraints and finite resources. Daly thinks economics must recognize reality and acknowledge that Earth's capacity to provide and endure is, in fact, limited and not infinite. People are now listening to him, even those who once derided him.

I think you should rethink economics and accounting. I urge you to think about EcoMetrics, to join the search for God's currency. Talk with Herman Daly. Change is coming. Change creates opportunity. A growing number of companies are beginning to think differently about their scorekeeping. It's just a matter of time (during the next "two-hundredths of a second") until all will have to. You could lead the way in this, and you should—for Earth's sake and for our grandchildren's sakes.

I could not tell from the immediate feedback whether I did, in fact, connect with that audience. There were a lot of stony faces; it would be unkind to describe them as blank. Afterward, there was just one request for a copy of my speech. I suppose time will tell; one never knows when a seed has taken root. The head of the firm did write me to say that my thoughts would not be ignored. I took great heart from that.

CHAPTER SIX

The Power of One

"Brighten the corner where you are." What if everybody did it? Most of the time, when I make an environmental speech, I'm preaching to the choir. Yet I am greatly encouraged and believe that the choir is growing, that the global brain *is* waking up. The number of "alarm clocks" to wake us is growing, too.

A quotation attributed to President Lyndon Johnson comes to mind. A rancher and landowner in his native Texas, Mr. Johnson was asked just how much land he wanted to own. He was said to have answered, "Well, just what I have and all that's next to it." Though Johnson's quote illustrated pure greed, in a filial way that's the choir we want singing the gospel of change: what's there now and all that's next to it. So to this swelling number, I continue to say that we are all part of the continuum of humanity and the web of life in general. We will have lived our brief span and either helped or hurt that continuum, that web, and the Earth that sustains all life. It's that simple. Which will it be? It's your call.

How can we help? I believe one person can make a difference. You can. I can. People coming together in organizations like yours and mine can make a big difference. Companies coming together—for example, customers and suppliers uniting in recycling efforts—can make a vast difference. Harnessing wind, current solar income, and hydrogen can make a monumental difference. Daniel Quinn's mission in his paradigm-shifting novel *Ishmael* is to change the minds of six billion people. If that happened and they decided to live their daily lives with Earth's welfare in mind, then the Earth, humanity, and all the continuum of life would indeed gain a new lease on life. The mid-course correction I think the Earth and humanity need probably depends on, more than any other one thing, changed minds—in other words, new paradigms. I have suggested one for business: doing well by doing good. But what will power this change?

In June 1996 Interface sponsored an event called The Power of One. We had seven speakers in the course of a day and a half together: David Brower, Bernadette Cozart, David Gottfried, Paul Hawken, Emily Miggins, John Picard, and Johnna Wenburg.

Each told us about his or her work. There was a wide variety. Hawken was chairman of The Natural Step USA and the philosophical heir apparent to Brower, an eighty-four-year-old ex-mountain-climber with a reverence for nature, who had been the country's most influential environmentalist for fifty years, dating from his days as executive director of the Sierra Club. Bernadette Cozart was greening Harlem and instilling beauty in a community, along with pride and self-respect in its people. Emily Miggins was saving trees and speaking up for women's rights in India and China, as well as right here at home. Johnna Wenburg told us about living with the orangutans in Borneo and keeping a chimpanzee orphanage in Africa. John Picard was making a difference in construction projects and in architectural and interior design circles all over this country. And David Gottfried had given up the real estate development and investment banking businesses to follow his heart and live out his love for the Earth by promoting green buildings.

The Power of One has become a recurring theme in our company, as many of our customers, as well as our people, recognize. That day and a half set me to thinking about the power of influence and the way all of those people were exerting influence that went far beyond the direct, immediate effects of what they were doing. Though we, as a company, have a very long way to go to sustainability, how far notwithstanding, I realize now that the journey is taking place on three levels: (1) the level of *understanding* sustainability; (2) the level of *achieving* sustainability; and (3) the level of *influence*. All this can be illustrated graphically; I use Interface merely as an example. (See The Sustainability Curve.)

The curve of *understanding* or knowledge—call it the learning curve—is not only about learning what and "where" sustainability is, but also about how to get there, including identification of the technologies, attitudes, and practices that are needed and how they should be developed. Getting well up this curve has brought us to the seven-front plan for climbing that mountain called Sustainability, as well as to the model for the sustainable enterprise.

The curve of *achievement*—call it the doing curve—plots the substantive progress toward sustainability. The gap between knowing and doing

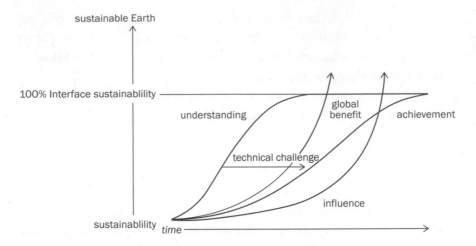

The Sustainability Curve.

represents the *technical challenge* (just not knowing how) or possibly the resource gap (not being able to afford it), but not, we hope, a gap of commitment or of willpower. In both understanding and achievement, we can only hope to approach 100 percent sustainability asymptotically; 100 percent is the limit. Getting well up the curve of understanding first, truly *getting it*, is imperative if the achieving part is to be intelligently directed; it's easy to take wrong turns and tangents such as downcycling precious molecules into less valuable forms—to get it wrong, even with the best of intentions. There are too many right things to do, and time is too short to waste resources on tangents. But understanding without those actions that lead to achievement is feel-good hypocrisy and mental self-gratification.

The third curve of *influence* is the one that will take our company beyond sustainable to restorative, putting back more than we ourselves take, and doing good to the Earth, not just no harm. The benefit to the Earth from inspiring others to take action, too, can be greater than from the company's own achievement. On the front of influence, it's worthwhile recognizing that, while there are limits to power from most sources, the power of influence has no limits (as my friend Dr. Zink reminds me regularly). There are no asymptotes, no mathematical limits, for influence. Yet influence collapses without the undergirding credibility that comes from actually doing it.

The fourth curve, the resultant curve that combines influence and achievement, may be labeled the *global benefit* curve. A company's total contribution is the sum of what it does and what it helps, inspires, or otherwise influences others to do. Like influence, global benefit also may be limitless; it knows no asymptotes, either.

Finally, it's worth pointing out that this same graphic representation can apply to an individual person as well as a corporate community of people. In fact, the graphic picture of an organization in many ways is the sum total of all its individuals' graphs.

Each of us individually is 1 in 5.8 billion (at this writing, but growing); yet all of us at Interface are 7,000 in 5.8 billion, more than one in a million. But Interface is one thirty-three-thousandth (1/33,000th) of the global economy. DuPont and BASF, both fifty times bigger, our major suppliers, are each one six-hundredth (1/600th)! All of us are resource-intensive, so together we have an even bigger impact than our numbers suggest. As a company, Interface can make an immense difference by setting an example, especially if we can influence DuPont and BASF to join our efforts—and Solutia (spun off from Monsanto), and Geon, and Shell, and all the rest of our suppliers. If our customers join in, too, we can make a colossal difference!

You (the reader), too, have influence. You have the Power of One. Your organization has influence—the collective influence of one and one and one. Knowledge, deep (not superficial) knowledge, getting well up that curve, comes first. Doing (taking action) must follow—in your personal lives and at work. Knowledge and action are critical. They give credibility and validity to your examples and to your influence, which can spread and grow without limit. You, too, can join in that positive feedback loop, doing well by doing good, a win-win for you and Earth.

I've loved Georgia Tech as long as I can remember. My fondest memory of childhood is sitting on my back steps on a crisp November afternoon, eating a pomegranate that had burst open from an early frost, and listening to the Georgia Tech football game on the radio. I remember Dinky Bowen kicking a field goal to beat Navy 17–14 in 1944. I remember Johnny McIntosh tackling the Navy fullback at the Tech goal line in 1945, forcing a fumble that a freshman named George Mathews picked out of the air to run ninety-five yards for a Tech touchdown. I also remember sitting on those same back steps a few years later on a Friday afternoon, polishing my football shoes before that night's game at West Point High School, talking with Lew Woodruff, an assistant to Coach Bobby Dodd, as he recruited me to come to Tech.

When the grant-in-aid (football scholarship) offers came from Auburn (Coach Shug Jordan), Georgia (Coach Wally Butts), Kentucky (Coach Bear Bryant), and Tech (Coach Dodd), there was never any question in my mind about where I would go. That shoulder injury my sophomore year ended my football career, and I was able to concentrate on my studies, work hard, do well, and graduate with my degree in industrial engineering.

Thirty-three years later, well into a career as founder and CEO of Interface, Inc., I found myself on the Georgia Tech Advisory Board (GTAB), a sounding board for the president of the institution on strategic issues. After six more years I became chairman of that board—at a most propitious time. It was the first year of new president Wayne Clough's tenure.

In his new role Dr. Clough set out to establish his own agenda for the institution, part of which was to engage the entire faculty and administration in a collegial effort to rethink and rewrite Georgia Tech's vision and mission statement. In due course a first draft materialized and, as chairman of GTAB, I received a copy.

To my dismay, the notion of sustainability could not be found in the statement, nor the words *environment* or *ecology*, even though Tech, with funding from the General Electric Foundation, had had a Center for Sustainable Technology for more than a year. (Talk about the Power of One and being in the right place at the right time!) Suffice it to say that the final version, the published version, commits Tech to work for a "sustainable society" in the lead paragraph and has other references to sustainability and the environment throughout. Sustainable technology has joined biotechnology and telecommunications technology as one of the three major thrusts, or strategic areas of study, the university is mounting for the twenty-first century.

I am also involved with my old school, now renamed the Industrial and Systems Engineering (ISyE) School and ranked number one in the nation by *U.S. News & World Report* for six consecutive years. Elsewhere in this book I have talked about the kinder, gentler technologies of the future, which I believe must emulate nature. Well, I've put my money where my mouth is and committed Interface and myself to endowing a chair within the ISyE School, the Anderson-Interface Chair in Natural Systems. I don't know of another such chair in any university. John Jarvis, the head of the ISyE School, has a standing challenge from me: to learn how a forest works. It's the work of a lifetime, but I believe as the forest's symbiotic relationships are understood, new organizing principles for industry will be revealed. Those, I believe, will shape the model for industrial systems in the twenty-first century and the next industrial era, and will define the means by which Interface, as the prototypical, sustainable company of the twenty-first century, will operate. I think Georgia Tech can and should lead in discovering this model. The timing is perfect. As Tech is planning a transition from the quarter system to the semester system, the curriculum must change, too. Biology will become important to engineers. That's an amazing change!

In a sweet irony, the friend who challenged my views and sent me the books that presented "the other side" has endowed a chair in another department of Georgia Tech: the Chair of Environmental Biology. The distinguished professors who occupy those two chairs are going to have a lot of fun together, and Georgia Tech's strategic focus in the twenty-first century, new curriculum and all, will be profoundly influenced.

It also pleases me enormously that Kalle Robèrt, having been introduced to the Georgia Tech community through my efforts and after brilliantly presenting the principles of The Natural Step to a campus gathering of some two hundred scientists and engineers, has experienced a surpassing acceptance at the institution by being invited to, and accepting, a position on the Dean's Advisory Board of the College of Science. The dean, Gary Schuster, was the self-pronounced chief skeptic of The Natural Step until he had had an hour-long one-on-one discussion with Kalle, a session from which I watched them walk away arm in arm after the most intense of discussions about the sustainability (or unsustainability) of nuclear power. I had the enviable privilege of being the proverbial fly on the wall. It was wonderfully enlightening! (Their conclusion about nuclear power? Yeah, maybe, *if* nothing went wrong. Who wants to risk it?)

One result of Kalle's influence is that Georgia Tech's Center for Sustainable Technology has embraced The Natural Step and its principles of sustainability. Another result is the creation of an interdisciplinary Sustainability Task Force to augment the efforts of the Center for Sustainable

Technology in sensitizing the entire campus. This task force's mandate extends at one level to influencing daily life on campus, and at another level to instilling the teaching of sustainability so it becomes ingrained in every course, and at still another level to research and outreach, to further the embracing of sustainability throughout society, especially Georgia-based industry. Good people are involved. Good things will result. Tech's role will continue to expand. Of this I am sure.

I look for Georgia Tech to be a force in America's climb toward sustainability, and the Center for Sustainable Technology, embracing The Natural Step, to be a powerful catalyst in generating that force.

CHAPTER SEVEN

To Love All the Children

G ary Schuster of Georgia Tech is the source of a most provocative story, which he said is true. He said that the scientists who developed chlorofluorocarbons (CFCs) to be refrigerants, propellants, and stain-resistant coatings knew they had invented inert compounds. They thought that the compounds, because they would not react with other elements, might last forever. And they knew the compounds would accumulate in the stratosphere. So they could see CFCs lasting forever, accumulating and accumulating in the stratosphere. Could such compounds be produced and marketed in good conscience? They actually raised the question!

But someone reasoned that no, the compounds, once they were in the stratosphere, would be attacked by ultraviolet radiation and break down into their basic elements, chlorine, fluorine, and carbon; so no, they would not last forever.

Unfortunately, no one asked the most important question of all: *And then what?* And then what, we all learned many millions of tons later, was that the free chlorine in the stratosphere would attack the stratospheric ozone layer, resulting in the serious weakening of the ozone shield, which protects life on Earth from deadly ultraviolet radiation. Eventually the unasked question was answered at great cost, not only to the producers of CFCs but to society as a whole.

And then what? is an important question.

So: And then what? Where does it go from here for Interface? Suppose Interface is very successful in its efforts toward sustainability? Suppose, further (for example), that global warming, to cite just one threat, turns out to be so vividly demonstrable and undeniably true that the whole world wakes up one day with a gigantic cry of alarm. Suppose consumer outrage erupts and markets shift overnight. Where will the capacity to respond be found? A lack of capacity for response could be a stupendous stumbling block to the Earth's welfare.

Without early and gradual tax shifts or tradable emission credits to effectively tax carbon and to allow markets to adjust in an orderly way and bring new technologies on stream, a repeat of the oil shock of the 1970s is highly likely, should the unhappy wake-up call come to an unprepared world.

At Interface we see that alarming possibility as an opportunity for which those with foresight should prepare. Thus, Interface has created One World Learning (OWL), a company within our company, whose mission is to be the repository for the wisdom and knowledge we are accumulating on this mountain-climbing expedition and to impart it to others. OWL's genesis can be traced to a brainstorming session to conceptualize a new, sustainable enterprise. We told a task force, "If you can dream up a sustainable business, we'll create it." In selecting OWL's name we have avoided using the word *teach* in recognition of the implicit wisdom of the Iroquois language which (I've been told) has thirty-five words for "learn," but not one single word for "teach." OWL is in the *learning assistance* business.

OWL is sharpening its skills internally, first by helping our people learn The Natural Step, using the pedagogy developed by Karl-Henrik Robèrt. OWL will help other companies, too, to learn The Natural Step, then move beyond it to other areas we are pioneering. For example, experiential learning, team building, personal breakthrough, and value shifting are facilitated by OWL under "Why?"—the program name we have adopted. We want Interface always to be a learning organization. OWL is integral to that commitment.

For another example, as a company we know a lot about merging cultures. As a river takes on the waters of its tributaries to grow wider and deeper, Interface has grown with some forty-eight acquisitions since 1982. There are many valuable lessons we have learned from those acquisitions and the companies we have acquired as they have been absorbed and hooked up. Further, the same methods for facilitating cultural synthesis have helped us open up our people, via the Why? experience, to the personal breakthrough changes that QUEST and EcoSense have needed our people to make, to become an empowered team. Command and control management would not have gotten the same results.

QUEST and EcoSense were part of a larger reinvention of Interface, Inc. No one and no company are born knowing these things. We hope to be in a position to help others who want to shortcut the learning process and soften the licks from the school of hard knocks. God knows we have taken a few along the way. And had some major triumphs, too—such as Maui.

Let me tell you about Maui. April 6, 1997, marked the twenty-fourth birthday of Interface and the beginning of our twenty-fifth year. A year or so before, I had casually mentioned to Charlie Eitel, then soon-to-be president and chief operating officer of Interface, that I had thought someday we might have a worldwide meeting of all our different subsidiary companies' sales forces, in one place, together. As our management team had cobbled Interface together over the years with those forty-eight acquisitions, we had left each company to conduct its own sales meetings—normally once a year in the winter months early in the new year.

With Charlie, who is naturally aggressive and who also has been exposed to the high rollers of the Young Presidents' Organization (YPO) (something I barely missed becoming a member of by not leading Interface to reach $1 million in annual sales before I reached age forty), you mention something like "worldwide sales meeting" but once. Before I knew it, Charlie had rented the Grand Wailea hotel and spa in Maui, Hawaii, for the entire week in which April 6, 1997, would fall. Why wait until the twenty-fifth anniversary? Why not kick off the twenty-fifth year with a big twenty-fourth birthday party, and celebrate all year long? Charlie knew the Grand Wailea from a YPO meeting. I had never seen it. But we had bought it for a week and had to figure out what to do with it. With uncanny foresight, Charlie began as soon as he had bought the week to urge the hotel management to think in environmentally sensitive ways—about recycling, for instance, a foretaste of what was to come.

In due course, with much planning, a theme emerged, "One World, One Family, A Celebration," and an objective for the meeting became clear: to take this far-flung company, so painstakingly assembled over the years to fulfill a grand design, and hook it up. A corporate strategy, *Diversify and integrate worldwide*, had seen *diversify* and *worldwide* executed much more effectively than *integrate*. Our meeting in Maui could complete that strategy and generate a shared sense of values at the core to provide the centripetal, binding force to coalesce this cobbled-together company into a unified whole.

Some subthemes emerged, too, the kinds of themes that, if developed properly during the week, could facilitate the realization of the primary objective: to hook it up. Among the subthemes were *people*, *product*, and *place*, the three P's on which we had focused as a company in the larger reinvention process that had started in 1993. I undertook to lead the development of *place*, the environmental thread, one of the three we would hope to weave into the fabric of the meeting throughout the week.

A chance to make a stepping-stone out of a stumbling block immediately presented itself. We also had reserved the Lake Placid Lodge in upstate New York for a three-day conference that had been planned for customers. But the sign-up had been so meager that we had decided to cancel the conference. With the hotel paid for and no conference to use it, we decided to salvage our cost and use the lodge for a planning retreat for the place team, which included our environmental consultants, the group I had dubbed the EcoDreamTeam.

At that time, September 1996, the DreamTeam consisted of John Picard, Paul Hawken, Daniel Quinn, David Brower, Bill McDonough, and Amory Lovins. This newly formed team (with John Knox filling in for David Brower, who couldn't be there) went to Lake Placid. Once there, the team learned for the first time where the meeting was to be, the Grand Wailea on Maui.

There was instant consternation, then revolt! Hawken's face clouded, and he asked, "Why there? That's the most extravagant, expensive, opulent hotel ever built. It would be totally inconsistent with Interface's efforts for the environment." In so many words, "Count me out!"

Quinn echoed the sentiment, and others joined in the chorus of protest.

Totally taken aback, never having seen the place, but thinking of the sunk cost, I stood my ground. "We are going," I said, "with or without an environmental theme."

Searching for a middle ground, someone said, "Maybe we could do something environmentally sensitive at the hotel."

Paul said, "That misses the point: the bigger issue how tourism is destroying the islands and the native culture."

I asked Paul whether he might be able to get a native-born speaker to address our meeting and enlighten us.

He said, "Yes, but you know what they'll say, don't you?"

"No," I answered, puzzled.

His stunning reply brought our tense discussion into perfect focus. "If you asked the native people of Maui, the indigenous Hawaiians, 'What can we do for you?' they would answer, 'Don't come.'"

It was a standoff, a tense one at that. But one suggestion saved the day: "Why don't we treat the Grand Wailea as a metaphor for how we live our lives? Let's make it a design problem. If we can change the hotel for the better, and use it as a classroom in the process, maybe we can all learn how to live our personal lives better." The Power of One, and what if everybody did it? Perfect! It was a brilliant idea. It saved the day and more. All the team members came on board and the ideas began to fly. The other linch-pin idea came soon after: Let's do something while we're there to love all the children.

The planning meeting was a great success. Now came the challenge of executing. Would the hotel cooperate? Would they allow us to change anything? What would we change if they would let us?

To get at all these issues, we organized a visit to Maui in November 1996. I used October to arrange to meet Takeshi Sekiguchi, the Japanese owner of the hotel, approaching the introduction in the Japanese way, through a Japanese friend, Mako Yasuda.

We hoped, through the efforts of our own Jim Hartzfeld, together with John Picard, Paul Hawken, and Bill Browning from Amory's orga-nization, the Rocky Mountain Institute, to eco-audit the hotel and gain some idea of the changes our people, as guests, could make to lessen our collective environmental impact, or footprint, while we were there in April. We hoped further that the hotel staff might be inspired, too (the influence curve). But we thought the chances were pretty slim without the owner's support.

Our first day there in November, Charlie Eitel and I sat down with Sekiguchi-*san*, his interpreter, and his American general manager, Greg Koestering. Immediately Sekiguchi-*san* understood our objective and, apparently, the opportunity. As soon as we told him what we wanted to do and suggested that his staff could learn with us in the "classroom," he turned to Greg and, without the interpreter's help, said, in perfect English, "Whatever they want, do it."

Wow! All the barriers came down; the caution vanished; the hesitancy evaporated. By the next day word had spread throughout the hotel staff. The day after, it was all over the island and demand began to build to let the islanders in on this happening, too.

The eco-audit proceeded swimmingly. The hotel, we learned, had comprehensive measuring systems for resource usage: electricity, water, propane, solid waste, detergent usage. The staff knew some of the problems, too: Suntan lotion in the swimming pool, for example, clogged the pool's filter system and increased electrical loads on the pumps.

The Grand Wailea was, to be sure, an energy hog, but it did provide an absolutely superb guest experience (thus its appeal to the YPO). In planning our changes with the hotel staff, we undertook not to institute changes that would diminish the guest experience.

With the eco-audit done and the impact-reducing ideas in hand, the planning accelerated as April approached. The number of attendees was coming into focus by now: eleven hundred, including one hundred people to be sent by our suppliers, almost a hundred support or resource personnel, and some nine hundred Interface associates from around the world (among them, fifty people from our factory floors, chosen in random drawings at each plant to represent their associates who were left at home keeping the plants running). But how to communicate with this international assemblage representing management, sales and marketing, and factory associates from thirty-four countries and speaking eight languages? How to sensitize them, get them on the same wavelength?

Hawken, Browning, and Hartzfeld set out to design an exercise to begin to do just that. They called it "The Global Village." In a group the size of ours, one person could represent about five million people of Earth. What an opportunity! To put eleven hundred people in a room representing the whole Earth, and illustrate the distribution of population, the maldistribution of resources, and misery in its numerous forms: hunger, disease, ignorance, want, mistreatment of women. We wanted a mind-shifting experience.

The crowd began to assemble in Maui on April 4, 1997. By April 5 everyone was there. A Polaroid photograph of each person was taken on arrival for a special purpose to be announced later. The first evening was an opening hospitality session—beautiful, smartly dressed people, buffet dinner on the hotel's luxurious grounds overlooking the Pacific—just what everyone would expect in such an opulent setting. A hint of things to come was offered as Jennifer Eitel, Charlie's daughter, led a local children's choir in concert for us. As they sang, "We are the world, we are the children," throats constricted a bit in the presence of such precious innocence. Still, it was your normal, glorious opening evening, but after it was over we collected and weighed all the wasted food—information for later.

The next morning, April 6, all eleven hundred participants assembled. The sheer magnitude of the undertaking to bring the group together brought goose bumps. Charlie recited one by one the names of all thirty-four countries where Interface people are located. Representatives from each country stood. The applause was enthusiastic, with the smallest contingents receiving the loudest welcome—demonstrating the spirit of support developed among our associates on the ropes courses of One World Learning and its forerunner, Pecos River Learning Center. Hook-it-up had begun there.

In my own opening remarks, I shared a startling fact, ascertained by Dianne Dillon-Ridgley, one of our directors: It was exactly one thousand days until the millennium. The stage was set.

The balance of the morning brought a series of inspirational speeches, highlighted by Terry Waite's jarring personal account of his imprisonment at the hands of his Lebanese captors for 1,763 days, most of the time blindfolded and in solitary confinement. Terry, a giant of a man physically, was powerful and moving, a testimony to the strength of a giant human spirit, too, standing on truth in the face of great adversity —the *people* theme of the meeting. And he provided the perfect, sobering preparation for what was to follow, the sensitizing session of The Global Village. Blind balladeer Ken Medema concluded the opening morning with a song improvised on the spot and just for the occasion. He summarized the extraordinary spirit that had already taken hold with "We'll Get It Right This Time"—a theme song for the next industrial revolution.

That afternoon, everyone took a seat in a U-shaped arrangement of chairs. In every seat had been placed a piece of paper containing some numbers. Hawken was master of ceremonies. He introduced the purpose, to give us all a better understanding of who we are, and the tragedy, the sorrow, as well as the celebration to be found in the human condition. We were instructed to stand as the numbers on the papers each of us had taken from our seats were called.

Bill Browning then described Spaceship Earth, a speck in the cosmos moving eight hundred times faster than a speeding bullet. He acquainted us with the scale of the solar system and the minuteness of Earth, especially what we know about its extreme contours, Everest and the Marianas Trench, represented by the thickness of a dime and a nickel, respectively, on a Dymaxion projection of the Earth laid out on the floor, 1:6,000,000 scale. Humble speck we are, indeed.

After I presented David Brower's "Six Days of Biblical Creation" to fix in people's minds the geologic time scale (some were hearing it for the first time), Paul proceeded to the exercise, beginning with a film that graphically demonstrated the population explosion from the beginning of the Common Era (CE 1) to present, and projected to the year 2020. It left everybody gasping, "My God!" as a map of Earth literally filled up with lights, each light representing a million people.

Various commentators had been selected to help orchestrate the exercise. The first request from a commentator came, "Would all who have the number one on your paper please stand."

The entire room stood. "You are the people of the Earth. Please be seated."

Paul explained that we were simulating Earth's human population, one per five million approximately.

The next commentator, "Would all who hold the number two stand."

Number two represented the 486 million people of Latin America; 93 people stood and were asked to remain standing. Then the United States and Canada: 56 people stood, representing 295 million, and remained standing. In like manner we progressed around the room to Russia and Eastern Europe (59 people, representing 309 million), Africa (140 people, representing 732 million), the Middle East (34 people, representing 176 million), Europe (80 people, representing 419 million), Indonesia (38 people, representing 201 million).

Then we came to Asia. A breathtaking crescendo began to build. Japan (24 people representing 126 million); Asia less Indonesia, Japan, China, and India (158 people, representing 830 million); then India, and 181 people stood, representing 950 million people! Finally China (everyone could see it coming), 232 people stood for 1 billion, 218 million. Whoosh! The collective gasp was audible, and nervous laughter punctuated the buzz. More than half the room represented Asia!

As the audience settled back in their seats, a commentator called the next number and 26 people stood, representing the 135 million babies to join humankind this year. Then another number, and ten people sat down, representing nearly fifty million who would leave us by dying. That left a net gain in population of eighty-five million, with sixteen people still standing.

Paul returned for the next revelation, looking ahead to the future. He set the stage again by referring to the population film shown earlier. "It took all of human history until the year 1800 for us to reach a population of one billion. One hundred years later we were two billion. About fifty years later, we reached three billion, and twenty-five years later, in 1975, we reached four billion. In 1985 we reached five billion. Today, 1997, we are at 5,771,000,000. The rate of population growth has slowed in the last few years; however, at current rates, how many more will there be in fifty years? Would number seventeen please stand." All eleven hundred people stood again. A doubling to twelve billion in just fifty years!

Further looks at the future: When number eighteen was called, 350 people stood, representing all the children with us today. We were reminded that *they are* the future. Then seventy stood, representing the elders over sixty-five years of age, who will be leaving soon.

Next we looked at the makeup of our population today, at what we do for a living: five stood, our soldiers. Seven stood, our teachers. One stood, our village's doctor. Eight stood, our forty million refugees. Our farmers stood next, 143 strong, 13 percent of our population, down from 40 percent in 1950. *All* workers, the 2.5 billion with jobs, were represented by 477 villagers standing. But more staggering was the response to the next number, those *without* jobs, looking for work and unable to find it: 190 villagers stood; almost one billion human beings, victims of population growth, uneven economic development, and labor-saving technologies. One could not help but think of the sixteen additional villagers (eighty-five million people) who join us *every year*, most being born into those areas where most of the one billion jobless already live.

The intensity grew as we began to examine the distribution of resources and industrial throughput: aluminum, iron, steel, chemicals, paper, timber—so disproportionate to the distribution of population. Income distribution, too, was shocking: 248 stood to represent those who earn less than $370 a year! At number twenty-eight, the richest one-fifth *and* the poorest one-fifth stood together, 440 in all. We learned that the richest 220 have sixty-one times the average means of the poorest 220. When the billionaires were asked to stand, no one stood. Paul had a tiny, two-ounce figurine in his hand at the podium. It represented, in proportion, he said, the three hundred billionaires on Earth who collectively receive as much income as three billion average Earthlings.

At number thirty, we began to examine the distribution of energy usage, then pollution, then food distribution. At number thirty-three, those who go to bed hungry each night stood—187, representing 981 million! Then 119 stood to represent the 595 million underweight and malnourished children—one in every three!

And so it went, grim fact after grim fact revealed. The plight of Earth, the vast inequities, and the vulnerability of the human species stood starkly exposed, all around. I found myself standing in the category of women in the world who have no access to birth control. I represented five million women, and I was joined by 183 (915 million) others. When eight people stood to represent the forty million of us who die from hunger or hunger-related illness each year, Paul reminded us of the simple arithmetic that equates this to three hundred jumbo jets crashing *every day*!

There were messages of hope, too. We learned that half the children in the world are now immunized against measles and polio, that life expectancy has increased by a third since 1960, that access to safe drinking water has almost doubled, that primary education enrollment is up nearly two-thirds since 1960, that food production is up in the last decade, that educational enrollment of girls has doubled over the last two decades, that infant mortality rates have halved since 1960, that three-fourths of Earth's people live in democratic regimes, that micro-lending institutions will soon serve one hundred million of the poorest of us, and that the usage of energy per unit of GDP is declining rapidly. Paul wrapped it up with Nelson Mandela's thoughts about individual responsibility and the power each one of us has to make changes happen in our world—Mandela's personal and unique version of the Power of One.

The Global Village experience was unsettling as well as inspiring. We didn't want to leave people in the grips of depression or despair. Dr. Zink had suggested a period of decompression and venting, so we concluded with small breakout groups. People talked with one another about their feelings for a while; then Dr. Zink, Larry Wilson of Pecos River Learning Center, and Archie Tew of One World Learning engaged the breakout groups to elicit comments. The comments came, thoughtful, articulate, representing the diversity of viewpoints in the groups. The comments of an African American male associate from Georgia epitomized the dominant impression of the day: "I had no idea before how my lifestyle impacted women and children I don't even know in faraway parts of the world. Now I know."

Everyone was engaged and, at the end, everyone was emotionally primed for the challenge. I told them about the Lake Placid experience and Hawken's comment: ". . . Don't come." And I expressed my wish that we be such a different group of visitors that we would be invited back gladly by the native people of Maui.

The challenge was issued: Do something to love all the children. Right there, forty teams of twenty-five or so people were formed. Native Hawaiians had been recruited in advance to become a part of each team, one per team. Each team was challenged to develop an idea of what we could leave behind as a permanent legacy for the children of Maui, not just an idea that would survive our leaving the island, but one that would still be there serving the children when we had all left this life. The natives were there to help the challenge teams make sure the ideas were relevant.

I then announced Interface's commitment to provide $50,000 to fund the implementation of the three best ideas, and the teams were asked to report their ideas three days hence. The teams gathered and went to work, but not before the Polaroid photographs from arrival day were explained. An entire wall of the hotel had been designated as a personal legacy wall. The photographs were displayed on the wall, and each person was invited to post personal commitments beside his or her picture throughout the week, indicating changes in behavior. At the end of the week the wall was full! More than seven hundred commitments to do those "thousands of little things" had been posted. The Power of One, a recurring theme the entire week, was manifest. Every commitment was preserved in the newsletter that was subsequently published to report on the meeting to all our associates left behind to tend shop and as a reminder for those who were there.

The next morning a representative of DuPont, one of the suppliers present, without prompting by anyone from Interface, asked to speak. He announced that the supplier group would match Interface's $50,000! As the teams began to report in, there were scattered, individual offers to contribute sums of money, $20, $50, even $100. So Paul tried an idea out on the group in plenary session. How many would give $20? A few hesitant hands went up, then quickly came down. How many, $50? No hands were raised! How many, $100? The whole room exploded, every hand up! In an instant the funding had mushroomed to $200,000 as every person there pledged $100 toward the legacy. The next day one team announced that it wished to withdraw from the competition for a share of the funds, and just sponsor its own project, adding to the total funding the cost of their project to be borne by themselves separately. Patagonia, a supplier of commemorative shirts for the meeting, pledged its profits on the shirts.

That same morning we announced the tactical objective of the meeting: to reduce our impact on the hotel every day during the time left, to make the hotel a classroom in conservation and sustainability. We gave the audience the consumption measurements from the first day: electricity, water, propane, solid waste (including that left from the buffet dinner on the grounds). Then, for the days left, ideas were suggested: Hang your towel and use it again tomorrow as you would at home; turn out all the lights when you leave the room; take shorter showers; use the glucose-based soap and shampoo that had been procured just for us (the fish could eat it!); sleep with your windows open and let the air conditioner rest; draw the blinds during the day to keep the sun out of the room when you're not there; take only what you really need to eat.

There had been a group of nearly identical size from another American company at the hotel the same week of 1996—a year earlier. That provided the perfect baseline with which to compare resource usage. Each morning Jim Hartzfeld of Interface Research Corporation and Bill Browning of the Rocky Mountain Institute reported on the comparisons for the day before, resource by resource. Usage began to decline as people pitched in. Even though the hotel staff had been thoroughly briefed and treated to lunch one day by Interface to "seal the deal," institutional inertia exerted its grip right away. Our people complained that the housekeepers wouldn't leave the used towels on the rack. The staff members were well trained, so change was difficult. They couldn't "untrain" themselves for this strange crowd. By the third day, though, they got it, and the resource usages declined more rapidly. One day water usage inexplicably spiked upward. What? Collective chagrin reigned! The entire group was so sensitized by now that a massive hunt for the leak ensued. No leak could be found. Devastation! Had our usage actually increased? As it turned out, the data had been incorrectly recorded. Relief! Whew! Water usage actually had gone down.

By the end of the week the resource usage stood at:

Water:	Down 48 percent, equivalent to the entire rainfall for a year on the hotel's 42-acre grounds.
Electricity:	Down 21 percent, carbon dioxide emissions reduced the equivalent of 42 acres of forest—trees freed to remove someone else's CO_2.
Propane:	Down 48 percent, equivalent to another 13 acres of forest; petrochemicals left for others.
Solid waste generated:	Down 34 percent, the landfill spared and $40,000 (annualized) in tipping fees avoided!

Overall, financial savings were $1,081,000 on an annualized basis! Pretty good for the hotel, but the Earth benefited even more. The Power of One, when everybody does it: Lesson learned! (There was a nice quid pro quo, too, when we received a $1 million carpet order from the hotel. How's that for doing well by doing good?)

The people of the laundry, with half their regular jobs to perform, planted an experimental garden on the grounds, using native species—under Bernadette Cozart's direction (who better?) and with the help of Interface Research Corporation's people. The idea: to figure out a better, more sustainable horticultural plan for the grounds and their semi-arid setting.

The sugar-based soap and shampoo were great; the sunblock provided, not so great. A lot of sunburn suggested a lot of determined, committed people who were going all out for Earth, which proved to be a bit too much for tender skin. They got lots of sympathy.

Throughout the week, one emotionally moving speaker followed another, with objective, unemotional presentations interspersed. Tom Crum, author and teacher, conducted a session on conflict resolution using aikido, a martial art in which he is expert, to illustrate and drive home his points—further weaving the *people* strand into the meeting. Paul Saffo spoke on the information technology revolution; Amory and Hunter Lovins, on energy efficiency. The EcoDreamTeam inspired everyone: Hawken, McDonough, Brower, Picard, the Lovinses, Browning, Cozart. Daniel Quinn, facing a deadline on his newest book, sent his wife, Rennie, in his stead. Ken Kragen, retained to help us visualize the plan for Maui week, inspired us with his account of organizing Hands Across America and We Are the World. The Hale-Bopp comet punctuated each day's end, as if it were there just for us, along with our own lunar eclipse. Just offshore, humpback whales breached and spouted in seeming approval.

There was world-class entertainment, too: John Denver, the Pointer Sisters, Kenny Loggins. John Denver composed and sang a new song just for us (perhaps his last original work), "Blue Water World." He obviously loved being there, and pronounced the week a turning point for sustainability. (We hope to cooperate with John's Windstar Foundation to use the tape of the Maui premiere of "Blue Water World" to honor John's memory.)

In the end, even such world-class entertainment, while extraordinary, became secondary. The environmental theme completely eclipsed everything else in its effect on our people. Without a doubt, eleven hundred lives were changed, and those eleven hundred will touch and influence many others.

The island people's interest in having a role, an interest that had begun to build during that first eco-audit trip, blossomed into a Maui Day event. Sponsored by Interface, the hotel, and a local group called Maui 2000, the event was staged on the hotel grounds under a tent. We expected a turnout of maybe fifty, but some four hundred people came from all five populated islands and from as far away as Tahiti and New Zealand. Indeed, Interface's influence had already spread throughout the region. The DreamTeam spoke, one after the other—the first time this inspiring, smart, visionary, and practical team of environmental leaders had ever appeared together on the same program. How appropriate that the descendants of a people and culture that lived sustainably for six hundred years in their island paradise, before we "discovered" them, should be the catalyst to bring together this modern-day equivalent of their own ancient wisdom, to inspire them to strive for sustainable living again. Maui 2000 lives on, reinvigorated and inspired by the DreamTeam.

And we have it all on tape, along with all the rest of the World Meeting. David Brower told a local interviewer that historians would look back on that day as a turning point for Earth. David said, "They [the historians] just don't know it yet." His remarks went unpublished, but someday some historian will be doing some research and David's statement will be there on our tape to be discovered, and perhaps (who can say?) the day will be marked.

The last day's program brought a moving response from the native Hawaiian community. Elders, led by seventy-year-old Kapuna Kealoha, and children came together to thank us, tearfully, both in their native language and in English, for the gift of Maui Day and to receive the Children's Legacy. For them, it was an unprecedented experience with *haoles* (foreign visitors). Their thank-you ceremony, in the native tradition, was touchingly sincere. Not one of the eleven hundred present was unmoved, and tears flowed freely. Culturally, past and present were united. Keith McLoughlin of DuPont, representing the suppliers, was adorned with a lei by one of the children. Keith had to bend over double to receive the tribute, so tiny was the tot. It's okay for a grown man to cry, especially at a time like that.

I was inducted into the family of Kekula Bray-Crawford, one of the community leaders who spoke to us. Kekula and I are sister, brother, aunt, uncle, mother, father, and cousins. I hope to learn what requirements that places on me when I can get back to Maui and see my sister-aunt-mother-cousin again. Of one thing I am certain: I, and any other Interface person, will be welcomed back. Meanwhile, Kekula tries to teach me by email. I am learning about this old, though new to me, culture very slowly.

Legacy ideas had poured in. The DreamTeam, charged with judging and selecting the best ideas, had thrown up their hands at the difficulty of their assignment. All forty ideas were good. How to choose? Hawken, in his inimitable and brilliant way, provided the solution: Let the Maui community, itself, choose.

Not so easy: The native community is, in fact, divided. There are factions. Some want to secede from the United States. Others want to modernize with the most advanced technologies. There are three main factions. Paul's idea was to require the factions to cooperate, each to appoint a representative to a council of elders, with the elders joined by representatives from the children of the community. A wonderfully reconciling structure, horizontal among the factions, vertical among the generations, children and elders brought together to think and act in cooperation and reconciling concert.

The Maui community accepted the invitation to form such a council to administer the legacy, and asked Paul and me to meet with them before we left the island. We agreed to do that, to hear their organizational plan.

The last evening brought to fulfillment what Charlie Eitel said was his initial vision for the World Meeting, dating from my first casual remark

More than a thousand participants at the Maui World Meeting join to form the Circle I of the Interface, Inc., corporate logo, hooking it up.

that set his thoughts and wheels in motion. It was, he said, a vision of a hooked-up Interface, all the people from those thirty-four countries, with key suppliers, and all the resource people who pulled the meeting together, dressed in white, all joined in a great endless circle, forming the Interface logo, the Circle I. It happened on a Maui golf course while a helicopter captured the moment. As the camera ran, a scattered crowd converged and coalesced, and the order of the Circle I emerged. It is a visual and emotional wonder, white-clad, waving people against the green backdrop of a carpet of manicured grass. The golf course with its pesticides and the helicopter with its noise were vestiges of the first industrial revolution for which we just could not find suitable substitutes. Next time, maybe.

Kenny Loggins sang for us that last evening under the stars. And Glenn Thomas sang, too. Glenn's song has special meaning, so let me tell you about it.

On a Tuesday morning in March 1996, I had talked about our environmental mission to the sales force of Bentley Mills, one of the Interface companies, during their annual sales meeting. I thought I had given a pretty good talk, but I couldn't be sure how it was received. People made nice comments, but then to me they would.

So when a few days later, over my email, totally out of the blue, came the following original poem from Glenn Thomas, it was one of the most encouraging moments of my life. It told me that at least one person in that Tuesday-morning audience (and I think he surely represented many more people) *really got it.* Here's what Glenn Thomas composed and sent:[1]

Tomorrow's Child

Without a name; an unseen face
and knowing not your time nor place
Tomorrow's Child, though yet unborn,
I met you first last Tuesday morn.

A wise friend introduced us two,
and through his shining point of view
I saw a day that you would see;
a day for you, but not for me.

Knowing you has changed my thinking,
for I never had an inkling
That perhaps the things I do
might someday, somehow, threaten you.

Tomorrow's Child, my daughter-son,
I'm afraid I've just begun
To think of you and of your good,
though always having known I should.

Begin I will to weigh the cost
of what I squander; what is lost
If ever I forget that you
will someday come to live here too.

I immediately emailed Glenn in reply that "Tomorrow's Child" just had to be set to music, and about two weeks later, an audiotape came. It was Glenn's own voice singing "Tomorrow's Child," words and music by Glenn Thomas. By the time we got to Maui, "Tomorrow's Child" was already becoming part of our corporate culture. Glenn sang his song for eleven hundred people that last evening.

Charlie's closing comments presented to the world a hooked-up Interface, and announced the World Meeting's objective accomplished! In closing I reminded our people that the Earth still needs a miracle. I offered the thought that they had been there, they had been the miracle, and now should go and live it—every day of their lives, at home, at work, and at play.

We concluded this magical week together, lying on our backs in the dark but for a starlit Hawaiian sky, listening to "Clair de Lune" performed by the Maui Symphony Orchestra. Chip DeGrace, who led the staff effort to organize the meeting, chose the piece because it had been the favorite of his father, who had died the year before.

Aglow with the miracle of Maui, we began to disperse the next day. Every participant took home a videotape of excerpts from the meeting, concluding with that hook-it-up formation taking shape. Nobody's sure how our media people, Light and Power, got the tapes made overnight for every single person.

But the Hawaiian council had asked Paul and me to meet, so we did. The main topics were how to organize and what to call the council. Their first suggestion, so sensitive, so polite, was to name the council for me. As sincerely as it was offered, I took it to be a gesture. In no event could I let that happen, and I declined as gracefully as I could, suggesting they devise a better name. They were ready, and in the most considerate way offered the name: Hoòkupu. In the Hawaiian language *hoòkupu* has two meanings: "a gift" and "the first seed sprouting out of the devastation of a lava flow." How appropriate! It reflected their basic plan to use the more than $200,000 pledged as seed money for a nonprofit organization that would, in time, with increasing funding, carry out every single idea that had been offered by the legacy teams, and more.

Searching for a way Interface could permanently be associated with the good that would flow from Hoòkupu, I told them about the origin of "Tomorrow's Child," and I recited the poem for them. The last piece fell into place as, at my request, the council adopted the name:

The Hoòkupu Trust
A Legacy for Tomorrow's Child

"Let's do something to love all the children"—that central vision—lives in Maui.

In a most delicious and astonishing irony, Jim Hartzfeld was on his computer days later and found a new meaning in *Hoòkupu*. He removed the punctuation and spaced the letters a bit differently, and there before his eyes appeared:

Hook up u

Mission accomplished? I'll say!

Synchronicity? Or serendipity? Who can say? Which was it when a young woman named Melissa Gildersleeve, working for the state of Washington's Department of Ecology, heard Paul Hawken speak, then bought and read his book? Which was it when she sent the book to her mother, Joyce LaValle, an Interface regional sales manager in Los Angeles who was pursuing a carpet order for a project called the ERC? Which was it when Joyce, fearing that she would lose the business because the environmental consultant for the ERC, named John Picard, had said that Interface just didn't "get it," sent the book to her boss, Gordon Whitener? Which was it when Gordon sent the book on to me at a most propitious moment, just when I was sweating over what to say to a new task force to give them an environmental vision, and was trying to decide what my new role should be in the company of my own creation *and* was trying to figure out just what John Picard meant when he said that Interface didn't "get it"? (And yes, what Interface itself should grow up to be?)

Yet as powerful, galvanizing, and culminating as the Maui experience was in providing the setting in which the environmental theme could take over the process of hooking up a worldwide company, there may yet be a role of even greater importance ahead for the model that emerged from Maui. *And then what?*

In November 1996 I was appointed to the President's Council on Sustainable Development (PCSD). The PCSD had been created by President Clinton in 1993 as a council of distinguished Americans from various backgrounds and viewpoints: businesspeople, environmentalists, labor leaders, government employees (federal, state, and local, including cabinet members), women's activists, and Native Americans. Charged by the president to reach consensus on policy recommendations pertaining to sustainability, the members with wildly divergent points of view reached an amazing degree of consensus during the first three years of the PCSD's charter. One result was a report, *Sustainable America: A New Consensus*.[2] It offers a wonderful vision of a sustainable society, one that would be difficult to fault from any point of view, especially given our starting point of where we are today, an unsustainable society.

Jonathan Lash, president of the World Resources Institute, a Washington, DC–based think tank, and Dave Buzzelli, vice president of Dow Chemical Corporation, co-chaired the council through its first three years, effectively putting sustainable development on our nation's agenda, raising public awareness of the numerous issues surrounding the concept, and leading the council to reach consensus about the environmental, social equity, and economic imperatives of sustainable development. It was a remarkable achievement.

But as President Clinton approached at the end of his first term, the PCSD wound down its activities, sacrificing precious momentum to the uncertainties of an election campaign. Then, with President Clinton's reelection, the council received a charter renewal and was charged by Vice President Gore to take on the task of implementing the vision outlined in *Sustainable America*, to continue to get the word out, raise public awareness, applaud and publicize success, and, in particular, take on the issue of global climate change. He asked for policy recommendations, again based on consensus, after (in the vice president's words) "looking long, thinking big, and being creative."

Many members from the first term have left the council, and others have been appointed in their places. Co-chair Dave Buzzelli has stepped down from that position but remains on the council, and I have been appointed co-chair to work alongside Jonathan, who continues— thank goodness.

Reluctant at first to accept Jonathan's bidding to allow him to recommend me to the White House for the co-chair, I finally relented, primarily because of something my friend Huey Johnson once said to me: "One policy is worth ten thousand programs." The Power of One, in a political sense.

I took the job and quickly found it to be a thankless task, one that I have likened to a very large sack of potatoes to be peeled—peel one and there are plenty more where that came from. But peel on we do, and the reconstituted council, after a year, begins to regain its momentum. Our first significant accomplishment was to deliver to the White House an agreed-on set of principles, a "this we believe" statement about global climate change. The principles are a good start, though in time we should (and, I think, will) go much further before we're finished, because the Earth needs us to. The principles are calculated to help President Clinton in climate treaty negotiations with other nations and, perhaps more important, with our own Senate to gain ratification of the Kyoto treaty.

Another objective, perhaps the one that can have the greatest value if we do it right, is to organize a National Town Meeting for a Sustainable America. We are planning the meeting for May 1999, hoping to make it a truly national, even international, event to stimulate a much-needed phenomenon: to get America to speak up for the Earth. For a network to "work," it first has to "net": to find each other and connect, to hook up (sound familiar?), so to speak. The people working on the meeting believe there is a vast potential network just waiting to "net" when people of like mind discover one another's existence and draw strength from their surprisingly large (we believe), like-minded numbers.

We are planning the conference with a center, Detroit, and with cities all over America connected by satellite television. Each major site will conduct its own concurrent local sustainability conference, alternately tuning into the center and showcasing and spotlighting local initiatives. In turn, each site will be downlinked to smaller cities, schools, universities, churches, other institutions, and homes all across America. We visualize a network of conferences to involve a new network of people, discovering one another, realizing they are not alone, and rising up with one voice to call for a sustainable society. If we're successful, business and government alike will have to listen. When the people lead, the leaders will follow.

Interface's Maui experience is the model for this renaissance, this monumental hook-it-up attempt that we hope will awaken in America the spirit and the heart of the next industrial revolution. We have a poster child for the conference, all set to go, "Tomorrow's Child." We have themes, too:

The Power of One. What if everybody did it? To love the children, all the children—today's and tomorrow's—of Earth. Perhaps Tomorrow's Child will become the poster child of the next industrial revolution, and Ken Medema's song, the theme.

PART TWO

Mission Zero and Beyond

by John A. Lanier

A Note from John Lanier

At the beginning of chapter 2, Ray called his story, the story you have just read, an epic. If I may be as presumptuous as he was, I would say that he had no idea how right he was. A full two decades have passed since this book was originally published in 1998, and his story has grown tremendously in stature and influence. Unfortunately, Ray is no longer here to continue as narrator.

On August 8, 2011, Ray closed his eyes for the last time. His family grieved, as did his friends, his employees, and all those who knew Ray or were inspired by his words and example. In time that grief transformed into fond remembrance and determination to keep climbing Mount Sustainability.

At the outset I want to say that I am rather intimidated by the task of assuming Ray's mantle as the teller of this story. Who am I to be entrusted with this task? The short answer: simply one of five lucky souls to be a grandchild of Ray. The longer answer: an environmental sustainability, circular-economy enthusiast whose family has empowered him to help advance Ray's legacy as the executive director of the foundation that bears his name.

Together with my family, the people of Interface, and many more sustainability champions, the foundation and the team of experts that advise it form an influential part of a broader movement of people and organizations who believe in creating a better future. Even while reflecting on the breadth of this movement, though, I find that Ray's absence is a gap. He was one of a kind, a radical industrialist who set a bar so high that it remains out of sight for most companies two decades later. But how much higher might Ray set his bar today?

The remainder of this book explores the ways in which Ray's vision—a future where business and industry heal the planet rather than harm it—not only endures but also evolves, becoming more urgent, increasingly more sophisticated, and ultimately more comprehensive. The prototypical company of the twenty-first century has not yet been fully realized, though I would argue that Interface is about as close to it as you will find. Once realized, this prototypical company cannot simply be a token of ingenuity and ethics. It must replicate and scale until it becomes the norm, and changes the entire economy in the process.

CHAPTER EIGHT

Nearing the Summit

It is always unfortunate when the star player of a sports team has to leave the game early with an injury, and doubly so if that game is for a championship. Sure, it makes it harder for the team to accomplish its goal, but the real shame is the look you see in the eye of the injured player—a desire to be out there helping the team win and a longing to be on the field at the moment of victory. That is how I imagine it with Ray's passing. I know how desperately he would want to be here for Interface's final push up this mountain that is taller than Everest. Reaching the top had been his dream for so long.

Indeed, Interface is close to accomplishing Ray's dream. It is nearly finished with its transformation from a linear, take-make-waste enterprise to one that operates on a cyclical premise. Soon the company will do no harm, and then it will work to *fulfill* Ray's dream by becoming a restorative enterprise. You will read more about that next journey in the chapters to come.

For now, though, let's revisit the original quest and the new dimensions it took on over time. After all, a lot has happened over the last twenty years, much of it under Ray's direct leadership. When Ray first wrote this book, the company's journey did not even have a name, but it does now. In 2006 Interface formally committed to Mission Zero, a pledge to ascend the seven faces of Ray's Mount Sustainability by the year 2020. Here, in a nutshell, is what that journey has looked like:

1. **Create zero waste.** The concept of waste does not exist in the natural world, and so a sustainable enterprise must similarly operate without waste. Interface is accomplishing this goal through resource efficiency, energy efficiency, operational efficiency, dematerialization, and recycling.

2. **Generate zero harmful emissions.** Whether through its smokestacks or effluent pipes, a sustainable enterprise cannot pollute the environment with toxic chemicals or climate-altering molecules. Interface is

accomplishing this goal through the same efficiencies listed in face one and by conducting rigorous life cycle assessments (LCAs), sourcing benign chemical substitutes, engaging in process redesign, and offsetting carbon emissions when necessary.

3. **Operate entirely on renewable energy.** Once again with nature as the model, a sustainable enterprise must meet its energy needs without relying upon fossil fuels or split atoms. Interface is accomplishing this goal through energy efficiency, directed biogas and landfill gas, on-site renewable electricity generation, and renewable energy credits (RECs) when necessary.

4. **Operate a closed-loop, circular manufacturing system.** Ecosystems constantly cycle materials, assembling molecules into complex structures and then naturally breaking them back down. A sustainable manufacturing enterprise must similarly rely upon renewable, biodegradable, and recyclable materials without extracting virgin, nonrenewable raw materials. Interface is accomplishing this goal through process redesign, innovations in manufacturing, sourcing renewable and biodegradable materials, and aggressive reverse supply chain efforts.

5. **Transport people and products as efficiently as possible.** Our modern transportation system is powered almost entirely by fossil fuels, so a sustainable enterprise must do all that it can to limit its impact on the environment through transportation of employees, supplies, and manufactured goods. Interface is accomplishing this goal by optimizing its supply chain, utilizing more efficient rail transport where available (avoiding shipping by air as much as possible), dematerialization, encouraging employees to use more environmentally friendly transportation options, and offsetting emissions when necessary.

6. **Sensitize stakeholders.** Just as no one species exists in isolation from others, no business operates in isolation. It depends on a supply chain and willing customers, and it influences and reacts to the actions of its competitors. A sustainable enterprise must encourage all of its stakeholders to join in the climb up Mount Sustainability. Interface is accomplishing this goal by being authentic in its pursuit of Mission Zero, telling its story as often as possible, sharing its learned lessons, utilizing a sustainability filter when purchasing materials from suppliers, encouraging carpet recycling (even of competitors' products), and asking employees to practice sustainability at home just as they do at work.

7. **Redesign commerce.** Ray called this face the "final ascent," a steep climb at the end of the journey where everyone must come together to reach the goal. Just imagine if the very structure of our economic system incentivized positive social and environmental outcomes, rather than negative ones! No single company or industry can ascend this face alone. Nonetheless, Interface stands ready to climb this face with others, and they work to accomplish this goal by redesigning their own businesses and seeking innovative new business practices that create true triple-bottom-line results.

People would often ask Ray how far up Mount Sustainability Interface had climbed, and Ray himself documented the early years of that journey in *Confessions of a Radical Industrialist,* a how-to manual for others who might dare to join him. But the characteristics of these seven faces make it extraordinarily difficult to answer that question. While the first five faces can be measured, the last two are different animals. How could we possibly know how many people Ray and Interface have inspired to become environmental mountain climbers? When has the company done enough in sensitizing its stakeholders? How on earth do you measure the fundamental redesign of the modern, free-market global economy? For that matter, how can Interface know when it has sufficiently redesigned itself?

Even within those first five faces, measuring progress is not always straightforward. As a global, billion-dollar company, Interface has bought and sold multiple subsidiary companies in the years since Ray's epiphany. With every acquisition, the company's environmental footprint increases. With every divestiture, environmental footprint decreases. Similarly, the company has grown and contracted right along with global economic booms and downturns. If Interface is selling more carpet, it will use more energy and material, and vice versa in a down-market. Measuring Interface's environmental progress is complicated by these purely market-based variations, which is why the company's metrics have evolved to measure impact on a per-unit-of-sales basis.

For example, consider the context of the fourth front, Closing the Loop. If Interface were a smaller company, it might be able to source 100 percent of its nylon face-fiber needs from post-consumer sources. As it stands, though, Interface is not able to get enough post-consumer nylon into its reverse supply chain to meet the demand for its products. There simply is not enough old nylon in carpet and other products coming back to the company, and sourcing through vendors who recycle post-consumer yarn is not sufficient. Should Interface sacrifice its goal of growing its market and increasing market share to ensure that 100 percent of its face fiber is always recycled content? I say no. Interface makes the most environmentally friendly carpet tile on the planet, especially when factoring in considerations like recyclability after use, carbon footprint, and water intensity. If Interface grows its market share, that means *more* carpet will be made at the highest environmental standard, even though that growth would decrease Interface's recycled content percentage since they cannot currently source enough recycled content. How do you properly frame metrics to account for this complexity?

There is another challenge in measuring progress when you are pioneering a completely new dashboard of metrics. "We move the yardstick all the time," said Dan Hendrix, retired CEO and current chair of the board at Interface. "When we started, no one thought about measuring carbon footprint. No one even knew how to measure our waste. We learned over time and moved the yardstick ahead, adding metrics to go with what was our latest understanding on where the yardstick should be to get us to our zero footprint goal."

What's the lesson here? It comes down to transparency and a willingness to constantly reevaluate progress through multiple lenses. Any company that sets out to climb Mount Sustainability must also commit to honest self-evaluation. You can never turn a blind eye to the indirect and latent ways in which your business damages environmental, community, and human well-being. Not only must you set the bar high, you must also be willing to move it even higher when the need arises.

Unfortunately, as the corporate sustainability movement has developed and even become mainstream, not all companies have approached it this way. Plenty of businesses have taken advantage of the lack of a universally accepted definition of the word *sustainability*, happily slapping the word *sustainable* or *green* on this product with 20 percent recycled content or that product with no toxic chemicals. While those accomplishments are better than nothing, they still run the risk of falling into the category of greenwashing. Sustainability should be an ethos, not a marketing tool. It certainly applies to products, but sustainability is even more about the organization and how it operates from a systems perspective. The reason that Ray's vision for Interface has had such far-reaching impact is precisely because it utilized a systems approach, understanding the company's many linkages to the Earth, society, and the marketplace. It was about so much more than just carpet.

Let us consider two examples that illustrate Dan's point about the importance of learning how to measure so you know where a zero footprint goal even is. You might have noticed that many of the seven faces touch on carbon footprint and greenhouse gas emissions, recognizing that business and industry are significant contributors to global warming. Interface is committed to lowering its carbon footprint—getting that to zero as well, in fact. Note, however, that the third front of Mount Sustainability does not say, *Be powered entirely by carbon-free energy*. The company chose to use the words *renewable energy* instead. The difference between the two is tremendous, as seen when considering the merits of nuclear power.

I will readily admit that, through the single lens of reversing global warming, electricity derived from nuclear power plants is vastly superior to electricity from the combustion of coal, natural gas, and oil. Nuclear power is carbon-free (or at least nearly so—some greenhouse gas emissions are associated with the extraction, refinement, transportation, infrastructure, and disposal of nuclear fuel). Those other fuel types certainly are not, with combustion of fossil fuels for electricity and heat being responsible for 25 percent of global greenhouse gas emissions.[1]

No one would accuse Interface of greenwashing if it had named 100 percent carbon-free energy as its goal for the third front. That is still an immensely high bar, and the challenge of global warming is of paramount importance. Had the company done so, it would have appeared to have scaled face three even more rapidly, as it could have included nuclear power's percentage of the company's grid mix in its measurement of progress.

Interface rightly set the bar higher than that, though. Remember The Natural Step that Ray described in chapter 4? Those scientifically derived principles of sustainability set the ground rules for determining what business practices are actually sustainable. How does nuclear power hold up when viewed in the light of The Natural Step?

Not terribly well, Interface determined. Nuclear waste remains radioactive, and therefore harmful to organic life-forms, for hundreds of thousands of years. That fact stands in direct conflict with the second and third principles of The Natural Step: that substances produced by society must not systematically increase in the biosphere, and the productivity and diversity of nature must not be systematically diminished. As a result, Interface does not treat nuclear power as progress on the third front of Mount Sustainability.

Here is a harder call: hydropower. On one hand, hydropower is just solar power captured in a different way. Solar energy evaporates water at the surface of the Earth; it then moves through the hydrologic cycle by condensing, falling as rain, concentrating in river basins, and eventually flowing to the sea. Humans have figured out that we can capture energy from rivers by damming them and running the flowing water through turbines. That electricity is renewable by any normal understanding of the term.

Unfortunately, a lot of the dams that generate hydropower come with a rather unfortunate side effect—they fundamentally change their local ecosystems. Damming a river will flood the surrounding terrain, displacing any non-aquatic life-forms in the region. Hydropower turbines present a threat to aquatic life, and the dams can disrupt migratory patterns of some aquatic species. Dams can also reduce a river's minimum water flow and, in the extreme, cause downstream portions of the river to dry up entirely.

Let's look at The Natural Step one more time. Ecosystem-disrupting hydropower does not violate the second principle like nuclear power does, but it certainly violates the third by harming the productivity and diversity of nature. As a result, Interface only counts low-impact hydropower in measuring its progress up the renewable energy face of Mount Sustainability.

Now do you see how difficult it is to determine true progress to a zero footprint goal? Interface cannot just scale all seven fronts of Mount Sustainability. It must also make sure that in doing so, it does not create unintended environmental harm. Even though considerations such as the health of local ecosystems and biodiversity loss are not specifically mentioned in the seven fronts, Interface still is mindful of them (and many others) in its climb up the mountain.

It is even harder to compare the progress of any two companies. How can we be sure that they are even using the same math? So long as pursuing sustainability is voluntary, we are unlikely to have a standard system of measurement. Greenwashing will remain a seductive trap for any company that does not "get it" the way that Ray and the people of Interface did.

A Conversation with Dan Hendrix

Dan Hendrix became CFO of Interface in 1985 and was named CEO in 2001. He succeeded Ray Anderson as chairman of the board in 2011 and retired as CEO in 2017.

JOHN: How did climbing Mount Sustainability change Interface?

DAN: Interface was always entrepreneurial, but the real accelerator to our business and to our mission was the culture that emerged because of sustainability over the last twenty years. We have been transformed as a company, in all of the best ways. We became a culture of dreamers and doers. The business case also emerged in ways we couldn't have imagined. We learned to look at returns from the perspective of a longer time horizon. Over time, investors began to appreciate that Interface would sacrifice nothing on the path to zero footprint. We were delivering superior products, designed with sustainability in mind; energizing the marketplace with our bold and authentic progress; dispelling myths about how "expensive" sustainability might be thanks to millions saved in waste avoidance; and creating a culture of courageous innovation. All of this remains true today.

How did the original tactical road map—the seven fronts of Mount Sustainability—fare over time?

In 1994 the first front, eliminating waste, immediately got eight thousand employees engaged in this sustainability thing as something they could actually get their hands on and get down to the shop floor. We put a reward system around it, created think groups, and put metrics around it. It was successful and became the springboard for the rest.

Our dashboard expanded as our understanding of sustainability expanded. For example, water and waste-to-landfill weren't part of original metrics. Measuring our carbon footprint wasn't even a part of it.

What were early wins, beyond waste elimination?

We had a big breakthrough when we discovered biomimicry. The idea that in nature there is no waste led us to innovate around dematerialization, while at the same time another part of our business was innovating around closed-loop recycling. We gained a lot of momentum quickly.

Ray managed to prove the business case for sustainability even when so many of the rules of the game weren't designed to reward positive environmental actors—especially back in the late 1990s. What part of that business case has surprised you the most?

We all underestimated the goodwill in the marketplace, especially from our key customers, the architects and designers who specify our product. Sustainability set us way apart with everyone—customers and suppliers. Early on, Ray was a voice in the wilderness, but at the same time he was so charismatic and so authentic that people wanted to follow him.

Let's talk about the fronts that are the hardest to measure—sensitizing stakeholders and redesigning commerce.

They go hand in hand, and it's really all about helping your customers, investors, and other stakeholders understand that you are disrupting the status quo in a way that will not only help them be more innovative and more intentional, but that will also take them outside of their comfort zone. It's a powerful notion, and we take the responsibility for bringing our stakeholders along on this journey very seriously. The true measure of our success is the fact that we have continued to grow and profit as a company, even through rough economic times, because we have a higher purpose.

When did you know it was real?

When I became the CEO in 2001. All of a sudden I wasn't just talking about the numbers, I was in the trenches with our people and they were so passionate; it was insane! It hit me in the face that we'd transformed this company. It was in our DNA, and I knew we weren't turning back, ever.

Ray always said that the status quo was a powerful opiate, but the risk in trying something new was worth the reward. Were there times when you felt like Interface was too far ahead and risking the business?

There were absolutely moments—when we were making investments, for example, for which the returns did not pencil—that I had to ask myself, "Do we actually back up so we can survive or do we have faith and go forward?" Sustainability was so embedded in our business that if we had cut it, we would have cut out the heart of our company.

What's next—not just for Interface but more broadly for authentic sustainability efforts? Will they continue to be viewed as good for business?

Millennials are demanding that companies do the right thing, and expect that business has a position on the circular economy, on the environment, and on social equity. The rise of social entrepreneurship, the increasingly global perspective that young people have because of social media—all of it will change expectations of business in big and in small ways. And today investors see climate change as a risk factor in a way that they didn't even five years ago.

The next Ray Anderson is going to be more balanced than we ever were about business.

So what is the role of the CEO in the innovation process today, with all of the externalities in business?

The CEO is responsible for establishing innovation as a priority, and for setting the expectation that innovation is not about business as usual—it's not about building a better mousetrap, it is about fundamentally rethinking the value proposition of your company and how you can deliver in new and increasingly more sustainable ways. It's also the responsibility of the CEO to make feedback a priority—listening to your customer, to the greater marketplace, and to people who don't support you is equally important. It's also about hiring for cultural fit. That's much easier said than done.

Beware the company that wants to look good, rather than be good. Ray's hopeful vision, the creation of the prototypical company of the twenty-first century, is completely dependent upon authenticity.

Fortunately for Interface, Ray was as authentic as it gets. "Ray was so charismatic with people and so authentic and believable that people wanted to follow him," said Hendrix. "He got it all right. He even got the restorative goal right, and he saw the emergence of the whole circular economy movement that is coming up. He nailed those back in 1995. Who would think in 1995 that society would be where it is today on these issues?"

That authenticity, that willingness to set the bar high and only move it up, now permeates the culture of the company. If you want a reason to trust the veracity of the metrics that Interface reports, I would point you there. What are those metrics, you ask?

Here you go:[2]

- 88 percent of energy used at manufacturing sites is from renewable sources.
- 100 percent of electricity used at manufacturing sites is from renewable sources.
- Energy efficiency at manufacturing sites has improved 43 percent since 1996.
- The average carbon footprint of Interface's carpet is down 66 percent since 1996.
- Greenhouse gas emissions' intensity at manufacturing sites is down 96 percent since 1996.
- Total waste to landfills from manufacturing sites is down 91 percent since 1996.
- 58 percent of raw materials used to make carpet are either recycled or bio-based.
- Total water intake intensity at manufacturing sites is down 88 percent since 1996.
- Thirteen million pounds of post-consumer carpet has been diverted from landfills through Interface's ReEntry program.

These numbers are tremendous! They also show that Interface has more work to do. Then again, Interface will always have more work to do. Beyond this point is true sustainability, and beyond that is restorative enterprise, and beyond that is being even more restorative.

In light of these numbers, let us revisit the "sustainability curve" that Ray introduced in chapter 6. When Ray first sketched this graph, Interface

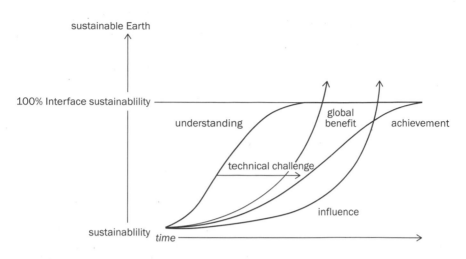

Ray's Original Sustainability Curve.

was still working its way up the *understanding* curve. For the most part, they've reached the top of this curve, as the company knows where it is in the journey and what else needs to be done. As for the *achievement* curve, Interface is right near the top. In other words, they are close to accomplishing Mission Zero.

That gap between the two, which Ray called the *technical challenge*, is still stubbornly present. Interface picked all the low-hanging sustainability fruit a long time ago when it eliminated obvious waste, made simple but effective energy- and water-efficiency retrofits, and incentivized its employees to find solutions to production-floor sustainability hurdles. In the middle years of the journey, Interface tackled bigger challenges like co-inventing machinery that separates carpet face from its backing, unlocking carpet recycling capabilities. The remaining fruit either is in the highest branches or has not yet ripened—in other words, the technical solution remains elusive or in the future. One example is the development of an abundant reverse supply chain for recycled nylon yarn. Another is finding a suitable replacement for the latex that keeps a carpet tile's face fiber stuck to its backing. A huge one is Interface's dependence on a fossil-fuel-based transportation system.

"It simply was not feasible to *not* put our products on trucks," said Dan Hendrix. "Figuring out transportation was always the hardest thing to do. I always said to Ray that transportation is one that we can't figure out ourselves. Maybe now, though, it can finally come about with companies like Tesla leading the way."

Dan is right. As a part of every sale Interface makes, a physical carpet tile has to be delivered to wherever the customer wants it. Ideally, those tiles would be transported on trucks that are powered completely by renewable electricity or hydrogen fuel cells, but unfortunately our transportation system does not work like that—yet. Talk about a technical challenge! It turns out that some technological innovations, like a fully sustainable transportation network, are beyond Interface's ability to control.

In light of these challenges, Interface often hears one particular question, usually asked with a fair degree of skepticism: Will Interface actually reach the top of Mount Sustainability in 2020?

Phrased differently, that question is about whether the technical challenge gap can be closed. Ray noted that the gap can be a combination of know-how or insufficient resources, but hopefully not a lack of commitment or willpower. I can assure you the remaining gaps have nothing to do with the commitment of the people at Interface. How do I know? Because the sustainability culture at Interface has persisted notwithstanding Ray's death.

Now, as you might imagine, losing Ray was difficult. Not a single square yard of Interface carpet had ever been made without Ray in the lead. Moreover, the man had been the head cheerleader for the sustainability climb for seventeen years. The company lost a fair degree of momentum with his passing. Would Interface possess the will, the ingenuity, and the passion to maintain its 2020 vision without Ray?

In many respects, the answer depends on Interface's new leadership. Recruited by Dan Hendrix and the Interface Board of Directors, Jay Gould left the CEO role at American Standard to become Interface's chief operating officer in January 2015. For two years Jay worked under Dan Hendrix, before stepping into the Interface CEO role himself. I asked Jay what had surprised him in the first few years of his time at Interface.

"Interface has changed me," he said. "I commonly call myself the accidental environmentalist because I didn't join the company for our commitment to sustainability. I did join because of Ray's influence on me and because I felt the company was a great platform for value creation. But being here has allowed me to open my mind and heart in a way that had not been done before."

Jay's first introduction to Interface was an interesting one. In the early 2000s he had been the chief innovation officer at Coca-Cola when he studied Interface to learn about its cultural success. Through the example of Ray and Interface, Jay became convinced that companies should embrace a higher purpose, and in doing so could become more valuable to the investor community and the marketplace. Jay brought this focus on purpose to every company with which he worked in the years to follow. While not specific to the environment, his commitment to purpose served him well when it came to being a leader prepared to guide Interface on the next leg of its journey.

As Dan Hendrix shared, "It's in our DNA. If Jay had tried to change the Interface culture, the company would have spit him out. Interface would have rejected him in my opinion."

With an intact sustainability culture, closing the technical challenge gap comes down to know-how and resources. The gap is real, but Interface knows exactly what is left to be done.

"We know what the gaps are because we know the components that are outside of our improvement," said Buddy Hay, assistant vice president for sustainable strategies at Interface. "For instance, we have our Mission Zero metrics, which don't measure improvement but measure what is left. At any point, we can go to any business unit and say, 'Our goal is zero, so what is left?' Sometimes we are talking about 3 percent and sometimes 10 percent, so that is measuring the opposite of achievement."

With a few more gains in renewable energy here and recycled content there, the gaps will shrink further and possibly be closed altogether. In a worst-case scenario, Interface can purchase various offsets to zero out any remaining negative impact, and then continue to work on eliminating reliance on offsets. Regardless, come 2020 I believe that Interface will have a legitimate claim to a concept that was once thought to be impossible—that a global, petroleum-intensive, industrial manufacturing enterprise can do no net harm to the biosphere.

Mission accomplished! I wish Ray could have lived to see the day. I fondly remember the passion and desire in his eyes when he spoke about his dream to see the view from the mountaintop. We are almost there, Ray, and I promise we will not pause too long to admire the view.

I get why some people might skeptically ask whether Interface will make it to the top of Mount Sustainability, but I hope their skepticism turns into inspiration and belief. We need more mountain climbers, skeptics

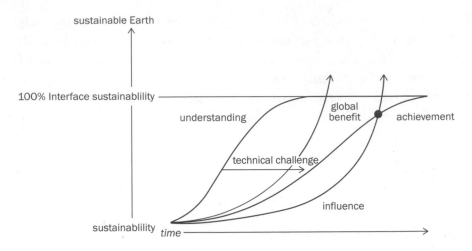

Where *Influence* Meets *Achievement*.

and "the choir" alike. We need more companies and individuals who are inspired, inquisitive, passionate, and creative thinkers who understand the intrinsic value in sustainability as a journey, not just a destination. We need them to see that Interface is an example of what is possible, as there is great power in possibility.

What is the significance of Interface's example? Asked another way, how powerful is the company's influence? Let us come back to that sustainability curve one more time.

Do you see that little black dot that I added? Something significant happens when Interface reaches that point on this graph. Prior to that intersection point, Interface's contribution to humanity, represented by the global benefit curve, is primarily the extent to which it has climbed Mount Sustainability (the water it has not used or the energy it has not consumed, for example). Beyond that intersection point, however, Interface's contribution to humanity is primarily the influence it has in getting others engaged in building a sustainable society. In other words, the value of Interface to the world is measured more by its influence than what it actually accomplishes in enhancing its environmental performance.

Now what if we assume that Interface is wildly successful at sending that influence curve up and up, higher and higher? What if it reached ten times higher than where the achievement curve plateaus, or a hundred or a thousand times higher?

In that case Interface's contribution to humanity from its influence will vastly dwarf its contribution from what it achieves. At a thousand to one, we could say that Interface's achievement hardly matters at all. What matters is the influence it has earned as a result of its achievement. In that case, does it even matter if Interface only makes it 99 percent of the way up Mount Sustainability?

Yes, it matters, but not because of the contribution of that last 1 percent. Ray said it himself in chapter 6: "Influence collapses without the undergirding credibility that comes from actually doing it."

Maximizing the impact of its influence is the true measure of Mission Zero's success. As the influence curve climbs, the journey becomes more successful, and Ray's legacy becomes more meaningful. Beyond that little black dot, Interface's story is less about doing no harm, and much more about being a *movement builder*. Just in time, too, because a sustainability movement (a real, ambitious sustainability movement) is exactly what we need to see transpire.

I have left one last question unanswered. If Mission Zero's ultimate success is dependent upon its influence, just how influential have Ray Anderson and Interface been?

"Ray's and Interface's impact was always way bigger than the impact a billion-dollar company can have," said Joel Makower, founder and executive editor of *GreenBiz*. "A Georgia Tech good ol' boy talking about sustainability and calling himself a plunderer of the Earth and reciting poetry and asking people to hug each other at the beginning of speeches? That affected and infected a lot of people. He wasn't the perfect man and Interface isn't the perfect company, but on balance he will long be remembered after *sustainability* is no longer a word we think about anymore."

Joel has as robust a perspective on the evolution of sustainability as anyone. He was writing about the environment well before Ray's epiphany and has continued to do so for decades. In a recent discussion, he shared an insight that I think is critically important: "One of the biggest challenges we face is we don't have any leaders—business, political, cultural—who can tell a compelling story of what happens if we get things right. We haven't had anyone to tell an irresistible story about what success looks like, on all levels from product to company to community to society to employees. In the business world, Ray certainly was the best teller of that story."

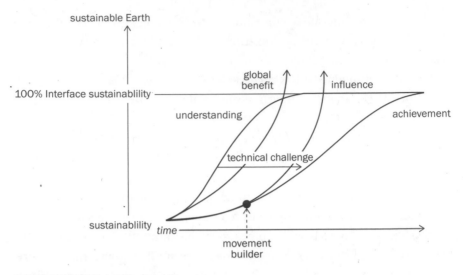

Interface Building a Movement.

Fortunately Ray told his story and told it well for many years. After hundreds of speeches and a 2009 TED Talk that has been viewed nearly one million times, just the telling of the story has created meaningful impact.[3]

This impact brings me to one final note about the sustainability curve. When Ray drew this graph, he did not yet know how influential he would become. I suppose out of his deep sense of humility, Ray drew the influence curve in such a way that it would cross with the achievement curve near the end of the company's approach toward 100 percent sustainability. I can say with confidence, though, that Ray misjudged. Ray and Interface long ago became true builders of the sustainability movement. I suppose that a more accurate curve would look like the above chart.

Interface is not alone in having its success defined by its influence. In a sense that is how we are each judged in our contributions to the future of humanity. For instance, I strongly believe that how my children lead their lives as a result of the values I teach them will have a much greater impact on the world than anything I do. If I teach them well, and they teach their children well, the influence will continue to grow. Call it "the power of compound influence," and it applies as much in the world of business and industry as it does in my home. The sooner we invest in building the sustainability movement, not just with our dollars but also with our hearts and minds, the more successful we will be.

A Conversation with Joel Makower

Joel Makower is chairman and executive editor of GreenBiz Group, Inc., creator of GreenBiz.com, and an expert on sustainable business and clean technology. He is the principal author of the annual State of Green Business *report and author of more than a dozen books on sustainable business.*

JOHN: How did you come to the environmental space and how did you come to know Ray?

JOEL: In 1989 I was drafted to produce the US edition of a British best seller called *The Green Consumer Guide* by John Elkington and Julia Hailes. The US version came out in early 1990, ahead of the Twentieth Anniversary of Earth Day, which was a big media moment. I was ordained as an expert on green consumer issues and quickly had a weekly syndicated column in ninety papers on this topic. I was preaching the gospel that every time you open your wallet you cast a vote for or against the environment.

But I quickly realized that there was no US green consumer movement. Also, the companies I was being asked to come in and talk to about consumer issues were themselves grappling with a whole bunch of things around energy, toxins, and waste. That was interesting to me and no one was writing about that. So in 1991 I started a monthly publication, *The Green Business Letter*—what we would now call a dead-tree, snail-mail newsletter.

I first heard about Ray from Paul Hawken, like so many people did. He told me he was having so much fun working for this carpet company in Atlanta and mentioned this guy called Ray. Then I met Ray and ended up spending time with him at various events, and talking with him both as a journalist and friend over the years.

Interface became the poster child for sustainable business, and I watched that journey for twenty-plus years, through the recession and all the things Ray went through when he came back as CEO.

What is your take on the scale and scope of Ray's influence?

Ray and Interface's impact was always way bigger than the impact a billion-dollar company can normally have because of Ray and his relentless enthusiasm and energy and willingness to be out there—a couple hundred times a year or more during his peak— beating the drum on this stuff. That impact had so much to do with who he was—this southern industrialist who relatively late in life had this epiphany. His persona became bigger than life. But it was also because people needed heroes, particularly in the corporate world.

I wrote a piece a year after Ray passed called "Why Aren't There More Ray Andersons?" When I would speak I often asked the audience to name a CEO of an industrial company who truly understands sustainability, has embedded it into their company at the DNA level, and is communicating that openly and relentlessly. A dozen hands would go up and people would shout, "Ray Anderson." And I would say, "Great. Now name another." Crickets.

Did Ray's influence have any downsides?

I once learned that your strengths maximized can become a liability, and I think there was a liability there in a couple of ways. There was certainly a liability in taking the eye off the ball at the company level, but beyond that I think it really had to do with that "name another" conundrum. There has to be another company out there doing this or it's not a movement. I think some people in the corporate world overdosed on Interface and Ray because he was the only person out there talking about this, which is not his fault.

It was a two-edged sword for sure. Ray may have been the only voice, but his was a critically important voice because it was something you could point to. And his was this great who-would-have-thought kind of voice.

On balance, he will be remembered long after *sustainability* is a word we don't even think about anymore. But someone on the business side really drove the conversation. It was Ray.

There was something else about Ray that I think was critically important. One of the biggest challenges we face is that we don't have any leaders—business, political, cultural—who can tell a compelling story of what happens if we get things right. We know what happens when we get things wrong from people who have told that story really well.

We haven't had anyone tell an irresistible story about what success looks like on all levels—from product to company to community to society to employees—and how this helps address not just environmental problems but also creates food, energy, housing and water security, and economic development and job creation and community resilience and well-being, all the way up to national security. In the business world Ray certainly was the best teller of that story. He didn't have the whole story because it has evolved along with technology and other capabilities, and the overall thinking has evolved. But he had a really good story, and I think that story—and of course the storyteller, with his backstory and his humanity and his great personality and southern drawl—enabled him to be the storyteller that we still so desperately need.

I fear that a lot of businesses fall into a greenwashing trap. They might start with good intentions but may realize an authentic pursuit of sustainability requires a lot more than they are willing to invest, and not just financially. What's your thought on that?

I don't believe there is that much greenwashing out there and I don't think greenwashing—defined as companies that are saying one thing and doing another—is a big problem. There are companies that may be beating their chest a little bit more than they deserve to. Is that greenwash or is that enthusiasm or hyperbole?

In fact most companies walk more than they talk for three reasons. One is that most of what companies do is about doing less bad—*This carpet has 43 percent fewer toxins than it did five years ago.* That is a great thing, but it still means we're 57 percent harmful. So doing less bad is a tough story to tell.

Number two, most of what companies are doing isn't part of the value proposition of what they sell. Most of GM's facilities around the world are zero waste to landfill, and that is a beautiful thing that is saving them millions if not billions of dollars. It's a huge lift and it delights their employees. But it has nothing to do with selling Chevys, so they are not going to message that in a Chevy showroom.

And that gets to number three. When companies talk about what they are doing right, they often illuminate problems they didn't even know they had. *How dare you talk about zero-waste factories when you sell internal-combustion engines?* It's easier to just do these things and reap the financial and employee and community benefits and not put a big label on it or run an ad or give a speech about it. It's just not worth it. When you stick your neck out, you can become a target.

You talk about that with almost any corporate sustainability leader and you'll get some aggressive head nodding. McDonald's would love to talk about what they do—their animal welfare policies and how they've gotten to zero waste or close to it. They have their own Mount Sustainability they are climbing fairly aggressively, such as moving the whole value chain toward sustainable beef. They can't talk about that stuff for the reasons I mentioned, yet they are doing it anyway and that is one thing I love: Companies that keep on it without necessarily talking about it. In fact, the former head of sustainability at McDonald's, Bob Langert, called it "green-muting," the opposite of greenwash.

So how do we create tomorrow's Ray Andersons?

My favorite questions in the world of sustainability all begin with the same four words: *What would it take?* And the rest of that question can involve anything from a job or career to planetary salvation. How you frame that question says a lot about who you are and what your ambition is, and how far you'll likely get down your path or up your mountain. It could be, *What would it take for my company to become a zero-waste company?* Or, *What would it take to create a circular model in the Yangtze River basin, where half of plastic waste originates?*

Those are really fascinating conversations—the kind Ray would love—and the good news is those conversations are happening in almost every sector all over the world. Some of them are just conversations, and some are conversations that end up in a white paper that sits on a digital shelf, but many become corporate strategies that take it to 2050 or beyond. They become another company's Mount Sustainability.

CHAPTER NINE

All or Nothing

I was eight years old when Ray had his epiphany. Being so young, I do not remember my grandfather before environmentalism became a driving force in his life. Even if I did, I doubt that it would have occurred to me that something significant had changed in his life. When I was a young boy, Ray was the person who took me to Georgia Tech football games and asked me what I was learning in school. I could not connect with him on matters of business, industrial ecology, environmental awareness, or natural systems.

Selfishly, I wish that his environmental epiphany could have waited until I was mature and sophisticated enough to understand it in real time (though for all of our sakes, I am glad it did not). Even more wistfully, I wish that he were still here today, to share what it feels like to get this close to realizing a dream as audacious as his. Alas, he is not, and I never did ask him certain questions on the front end. Questions like, "How long do you think it will take?" and "What will be the hardest part?" Perhaps I would have asked him, "Are you scared?"

A particularly interesting question would have been, "When you succeed in making Interface truly sustainable, what do you think will be your most important lessons learned along the way?" I realize that the question would have been impossible to answer back in 1994. No industrial enterprise had ever pursued environmental sustainability. It was *terra nova*, with no model to follow. I would have enjoyed hearing Ray's predictions of the future.

With the benefit of hindsight, it is easy to point to crucial learnings. This chapter will cover four, but before exploring them, I want to make two points. The first is to say that this chapter is not a playbook of *what* to do to green a business. A number of excellent authors have written those books, and I would recommend ones like *Green to Gold* by Daniel Esty and Andrew Winston to anyone wanting to know the concrete steps to follow in pursuing sustainability in their workplace.

Rather, these are the lessons that show the *why* behind the *how*—instructional observations and contributions from many who have touched the process, to help us understand why Interface has thrived in its journey up Mount Sustainability. While not a how-to, these insights can be applied to any enterprise that endeavors to "do well by doing good," as Ray was fond of saying. Because of these learnings, Interface has shown that the prototypical company of the twenty-first century will be better positioned not just environmentally and socially, but economically as well.

Second, these lessons do not exist in isolation, but rather arise from Interface's systems dynamics. They are inextricably linked. The goodwill of the marketplace helped drive Interface's sustainability innovations, and vice versa. Because of the company's culture of engagement, waste and efficiency gains were discovered. Finding those gains then reinforced employee engagement. It's like a sapling growing into a tree. That tree develops wood, bark, sap, leaves, and roots, and there is no alternative reality in which that tree matures without one of those components. It is all or nothing.

As a result, if a business wants to enjoy all of the benefits of a sustainability commitment, that business must commit holistically. Waste reduction efforts or renewable energy initiatives will underperform if employees do not commit to a higher corporate purpose. Innovations in products and processes will not revolutionize a business without bold and clear sustainability goals. Unless a business's leadership embraces an authentic sustainability ethos and encourages all employees to do the same, that business will find no more than middling success in any of its sustainability initiatives.

Lesson Number One: Sustainability Can Be Financed Through Waste Reductions and Efficiency Gains

As any introductory accounting student knows, profit equals gross revenue less costs. For profit to increase, either revenue must increase or costs must decrease. The mathematics are unyielding.

For this reason, businesses have sought to reduce costs since the beginning of our modern economic system. Operating costs, costs of goods sold, capital expenditures, interest payable, taxes, or any others: The type does not matter. In each case, less is more.

That said, costs are necessary. Without factories, no products are made. Without salespeople, inventory is not sold. Without taxes, we do not have a functioning society within which business can operate.

Waste is a bit different, however. Wasted energy, water, or raw materials have no silver lining. Waste is a drain on a system, and its reduction is always beneficial. Natural systems have no waste for a reason: It serves no purpose.

The sentiment was captured well by Joel Makower in *The New Grand Strategy*, cowritten with Mark Mykleby and Patrick Doherty: "At its core, resource productivity is about reducing or eliminating waste in industrial systems. Waste has long been the bane of manufacturing; it represents inefficiency and lost profit."[1]

How do we define *waste*, though? Some businesses might use the term solely when referring to waste that is hauled off and dumped in a landfill. That concept does not account for excess water and energy usage, however. When we're talking about water and energy efficiency, when do we know that we are as efficient as we can possibly be? It is only after a business improves its efficiency in these categories that it can say it had been wasteful previously. Ultimately, there is a physical limit to efficiency gains, but those limits are not precisely known.

We should also acknowledge operational waste. The incorrectly priced invoice and the box that was delivered to the wrong location certainly constitute waste. Leave payments to an employee injured on the job are waste; doubly so on account of the harm to the person who is injured. All of this shows that a broad definition of waste is important. That is why Ray defined it in chapter 1 as "any cost that goes into our product that does not produce value for our customers."

When viewed this expansively, waste reduction becomes a vast opportunity for improvement, with savings to be found all over the place. The savings are rather surprising, actually. Classical economic thinking would say that any waste reduction effort with a marginal benefit exceeding marginal cost would have already been taken, leveling off at an equilibrium point. This sort of thinking ignores the fact that real people must actually be looking to reduce waste in order to capitalize on the opportunity. That is the whole point of QUEST, which Ray talked about in chapter 1.

For example, one associate in Georgia suggested replacing massive beams of yarn with smaller, movable creels. By doing so, yarn could be more precisely arranged so that a run of the machines would not have as much excess yarn left over. His idea cut yarn overruns by 54 percent.

An engineer was looking for ways to reduce water usage at one facility. He suggested a brass nozzle be used to control water flow. That one tiny change saved two million gallons of water per year, worth thousands of dollars.

At a factory in Northern Ireland, an associate determined that the width of backing material during production could be shortened from 220 centimeters to 218 with a slight modification. That tiny reduction equated to tens of thousands of square yards of saved material.

Sounds easy, right? Just tell people to eliminate waste and count the savings as they roll in! Well, not quite. It turns out that employees need motivation. They must also understand why waste reduction is important. Even at Interface, it was slow in the beginning, as its people did not necessarily see waste in terms of opportunity.

"It was frustrating for people in manufacturing and cost accounting to say, 'We don't accept waste in our factories,'" said Dan Hendrix. "Theoretically it sounds great, but practically it is a lot harder. So we put a reward system around it and gave bonuses to people. Even 10 percent of my bonus was tied to our QUEST waste reduction."

Aha! Waste reduction can be an opportunity for not only a business, but also its employees individually. By aligning incentives company-wide (and company-deep, from the CEO to the factory worker), waste became the Easter egg that everyone hunted. With this alignment, waste reduction ideas poured in every year, all in pursuit of 10 percent annual reductions in waste. People truly bought in, and they have stayed that way. On this topic, I asked Buddy Hay if he thought Interface would make it as far as it has: "No way, even if you go back as far as 2000," he replied. "We got here because no one ever stopped. The first question is always, 'How much is it going to help us,' and 'How much does it cost' comes next. How much closer is it going to get us? You never get in trouble for saying I can move the needle."

In chapter 1 Ray said that waste reductions in the first three and a half years saved $67 million. Bringing that number up to speed is difficult for a variety of reasons, but I can conservatively say that savings have totaled in the hundreds of millions of dollars. Ray was fond of saying that cost savings in the waste reduction front alone paid for the whole sustainability effort.

So let us consider the cost savings in terms of what they have enabled. Here is Buddy Hay once again:

> *We knew we couldn't afford renewable energy until we got energy to its irreducible minimum. We were spending $11 million on energy, so our goal was a 50 percent improvement in energy efficiency. We knew if we could improve that, we could pay for the renewable energy pieces. Today we are still using the same philosophy, and I guarantee our energy efficiency improvement of 43 percent since 1996 has more than paid for the renewable energy that we use today. Similarly, dematerialization has paid for the premium that we pay for recycled content in our materials.*

Some businesses might have simply pocketed the efficiency savings without worrying about renewable energy or recycled content. But that didn't happen at Interface, thanks again to a whole-systems approach to sustainability. If employees had seen Interface become greener on the waste side without pursuing renewable energy and a closed-loop system, they could have felt that it was all about the monetary savings and not the higher purpose. Under those circumstances, would employees have been as motivated to participate in QUEST? The end-of-year bonus check only motivates a person so much. Moreover, would the marketplace have looked on Interface as favorably for its efforts? Classical economics has it wrong once again—the marginal cost/benefit analysis of waste reduction would not account for employee engagement and the company's external reputation.

A focus on waste reduction and efficiency gains has another, more subtle benefit for a business. More often than not, corporations are price takers instead of price makers with respect to their supply chains. In other words, most businesses do not represent enough macroeconomic demand to influence the market price of their material and energy inputs. A barrel of oil costs whatever it is going to cost, and the same with a board foot of lumber or a bushel of wheat, regardless of who is buying. Corporations simply take the price in the market; if their supply chain prices suddenly spike, they must roll with the punch.

The more resource-efficient a company is, the better equipped it is to handle supply chain fluctuations outside its control. In this sense, driving waste to zero is a risk management strategy. In the case of Interface, the company makes a product that is petroleum-based. Oil prices have spiked in the past, and they likely will again. By being so energy- and resource-efficient, and by creating closed-loop industrial processes that recover old carpet, Interface is well positioned to remain competitive when oil increases in price.

Beyond simple risk management, this creates a competitive advantage! Interface's competitors have certainly improved their environmental performance since Ray had his epiphany, but none have been as successful in increasing energy and material efficiency. Interface is well positioned should oil ever reach $150 per barrel, and in the meantime it continues to enjoy the benefits of its efforts.

So why has Interface benefited from its waste reduction efforts? It aligned its employees' financial and cultural interests with its own goal of becoming more resource-efficient. It then was willing to invest the financial gains from this efficiency in other sustainability initiatives, proving internally and externally that the company was authentically committed to sustainability. Many of these investments generated further efficiency gains in the long run. Finally, with its efficiency gains, Interface is now well positioned for supply chain variations that are likely ahead in the coming years.

Lesson Number Two: Sustainability Is an Accelerant to Innovation

"How can we get rid of the glue?" That was the original question posed in 2006. More than a decade into its journey, Interface was fully aware of how nasty the glue was that traditionally adhered carpet tiles to the floor. After being slathered all over, the glue would off-gas volatile organic compounds (VOCs) that contributed to poor indoor air quality. Some sort of adhesive was necessary, though; otherwise the carpet could buckle as people and things moved across it. Interface needed a replacement.

The innovation team at Interface got to work, admittedly with a false start. "The idea was to make a glueless system, and the engineers came back with a locking/fraying system," said John Bradford, Interface's chief innovations officer. "It took three engineers to put together and was just wild and crazy."

Glue is easy to use, even if harmful. The innovation team's first solution was far more complex, and installers would not respond well to a product that would make their jobs harder. Not only would the solution need to eliminate glue, but it would also need to be as user-friendly as glue has been.

Framed that way, getting rid of glue was a sustainability goal that constrained Interface's engineers. I am reminded of the scene in *Apollo 13* when the crew starts breathing too much carbon dioxide, and the engineers back home have to jerry-rig a carbon dioxide filter that the crew could then assemble on board their spacecraft. Ed Harris, playing the lead flight director, said, "Well, I suggest you gentlemen invent a way to put a square peg in a round hole." He meant it literally.

On board that spaceship, the crew faced unbending constraints. They could not stop by a convenience store for a new part. Conversely, innovation in pursuit of sustainability does not have to be a limiting factor, even though it might seem that way at first. Sure, reducing an environmental footprint is about minimizing a negative impact, but the innovation solution does not require you to do the same thing, just more minimally. As Bradford told me, "Sustainability lays a constraint in the middle of your design that causes you to change your mind and find a more elegant solution."

So toxic glue was a constraint in the middle of the design. What was the more elegant solution? How did Interface step back and allow their engineers the freedom to design a better way? They asked a better question.

"How does nature adhere?" Well, that is a different question, one full of possibility rather than limitation. It is a question that invites exploration rather than reductionism. It is also a question that suggests humanity has a lot to learn from the natural world, a concept at the core of the design discipline called biomimicry.

Janine Benyus coined that term in her book *Biomimicry: Innovation Inspired by Nature*, and for years she served on Ray's EcoDreamTeam at Interface. She summarized biomimicry as "a process of understanding function, finding design models in nature, emulating those models, and then evaluating them." It was the perfect approach for the Interface engineers who were trying to replace glue.

A group of people were gathered around two tables exploring how nature adheres. One table was looking at the gecko, which is able to climb walls and even hang upside down because of the microscopic hairs on its feet. Those hairs take advantage of a phenomenon called Van der Waals forces, which are attractive forces at a molecular level. This team realized that nature often uses natural forces to adhere, and that one constant and helpful force in keeping carpet on the floor is gravity.

A Conversation with Janine Benyus

Janine Benyus is a biologist, author, innovation consultant, self-proclaimed nature nerd, and member of Ray Anderson's DreamTeam. She popularized the concept of biomimicry with her 1997 book Biomimicry: Innovation Inspired by Nature.

JOHN: When did you first meet Ray and how were you introduced?

JANINE: It was 1998, right after *Biomimicry* came out. Paul [Hawken] said, "You need to get involved with this carpet company. They are way more than a carpet company."

I met Ray at a face-paced DreamTeam meeting that made me feel like I was hurtling at Mach speed, blown back by the g-force of a language and industry culture that was foreign to me. "Why am I here?" I kept saying. "What is a biologist doing here?" Ray was pretty quiet throughout the meeting, and then with his hawklike curiosity and focus, he looks over at me and says, "How would nature make a carpet?"

I said, "It would probably start out as liquid, be poured into the room, and would self-assemble with all the functionality that you want and absolutely no waste and no manufacturing plant . . . like a beetle's shell." He tilted his head, and I could tell his wheels were turning. What? He kidded me about it a little and I said, "It sounds odd, but that's how it works in nature. Seashells self-assemble from materials in the ocean, composites like wood and bone and spider silk—they all self-assemble. Nature's chemistry builds from the bottom up—molecules attract and puzzle together from a watery solution with very little energy input. I honestly think that is how it would go." He kept at it, asking more clarifying questions, and then, later, "How would a company function like a forest?"

It was just like him to move from a technical question to a societal one. Ray was a very scholarly guy and did a lot of reading. He also had this clear moral code. Carpet was just a vehicle for changing and transforming the industry.

What is the process of biomimicry? What does a good biomimic do if they are seeking innovation at their company for a product or a process?

The first step is to understand what you want to accomplish. Biomimicry is about function. It's not what you want to design; it's what do you want your design to do. Function is the bridge that allows us to find models in the natural world that have already solved what we're trying to solve. If Sherwin-Williams comes to us and asks, *How do we get the cadmium out of yellow paint?* We might ask, *What is it that cadmium does in the paint—what's its function?* Then we could search and find out how other organisms accomplish that. But real transformation goes even deeper. It happens when we ask, *Why do you want to create a nontoxic paint*—what is it that you want the paint to do? *We want to create the color yellow.* Aha! You want to signal the color yellow. Now we can open up the solutions space.

Function lets us "biologize" the question so we can find it in the biological literature. When we ask, *How do organisms create or signal the color yellow?* we find all kinds of ways that yellow is produced. Whether it is produced structurally or by non-toxic pigment or through luminescence. Lots of ways to create color, and none of them use cadmium! This is where radical new products and services are born.

To find the best natural models, we do an "Amoeba through Zebra" report, searching across the animal kingdom from bacteria to fungi to plants to animals. We look across taxa because the mechanisms and chemistries are very different. Once we've assembled our catalogue of life's chemical, physical and behavioral strategies, we look for commonalities, and pull out common design principles. In the case of a structural color, it might be, *Yellow is often created with a certain number of layers of a certain thickness to refract light.* It doesn't matter what material is used, or whether it's a moth or a peacock or a butterfly, this structural color recipe comes up over and over again. That's our design principle, and it can be emulated, perhaps in a non-toxic, light-permeable "paint."

Now we get to work with our clients to emulate this new approach in the most biomimetic way.

This emulation phase is very iterative because other questions come up: What material should we make it out of, or how should we package and ship it? It's not a linear thing. We're constantly thinking about the characteristics we want this product to have. What operating conditions will it have to face? In the natural world, an innovation (we call them adaptations) is one that is well adapted to context. Does the product have to perform well in the cold, in rain, in saltwater? Should it be able to repair itself? Is it better sold as a service? How will it evolve? Does it need to operate near living things? If so, it needs to be life-friendly.

That's what's great about biomimicry—you are looking at solutions that are high performing and life-friendly. Otherwise they wouldn't have lasted on Earth for so long. We've created a list of characteristics that virtually all organisms and ecosystems embody, and we call them Life's Principles. We use this checklist to make sure we're creating something, from the start, that is life-friendly, energy-efficient, sparing in material use, evolvable, and manufactured in safe ways. We use it to scope a project and then evaluate our ideas. This helps us do biomimicry at a deeper level, beyond a shallow mimicry of form.

What was the first real project that you did for Interface?
Biologists at the design table began with David Oakey [Interface product designer]. David had read the book, seen me speak, and became a real fan of biomimicry. Long before we had the Biomimicry Institute, we would call together people who were interested in the field and talk about what to do next, and what the field needs. David was one of the people who would come to those early meetings, and then he decided to host a workshop. He and his designers walked outside with Dayna Baumeister [cofounder of Biomimicry 3.8] and asked, "How would nature make a floor covering?" They started to look at leaves and rocks of the forest floor—very simplistic—but what came out of it was

one of these insights that is hard to describe, but it happens when you look at many instances of something and you try to figure out what the commonality, or deep design principle, is. That's how biomimicry works.

What the designers started to notice is when you pick up a leaf and look back at the ground, you can hardly see anything changing. They also noticed that because of the random order in the natural world and the fractal nature of things, the colors always went well together. It was this idea of random yet harmonious patterns that to them was really a breakthrough because at that time they were creating a precise, elaborate pattern on a roll of carpet and then cutting it into squares. You had to match the squares up to recreate the pattern on the floor, and if you picked up a square that was soiled and replaced it, the new one stood out in a sore-thumb effect.

A lot of waste was involved in cutting carpet to pattern, which turned out to be a pretty big deal. Yet when you look down at those leaves, no two are the same. There is harmony yet individuality; it works. So David had an idea: What if we made all of the carpet tiles different and made a harmonious color palette? And that became Entropy.

It took David Oakey being brilliant about the design and color choices and the people who programmed the computers to make it all work—no small feat. After that, David put Dayna on retainer and said, "Please be my biologist at the design table." Interface ran "Out of the Box" meetings each quarter with Biomimicry Guild staff, and they would kick around a question such as *How does nature adhere?* Other breakthroughs came out of those sessions—product innovation like Tactiles and even organizational innovations. These days, we're working to answer Ray's second question to me—*How would a company function like a forest?* We're challenging Interface factories and office buildings to produce the same level of ecosystem services as a high-performing native ecosystem—we call it "Factory as a Forest."

Ray's foundation supports the Biomimicry Global Design Challenge in hopes that biomimicry becomes as commonly known as any technology we can hold in our hand. Our goal is to see it normalized. Do you share that goal?

One of the theories of change we have is that biomimicry is best taught through being able to point at a nature-inspired innovation and say, *Oh if that's what biomimicry can do, shouldn't every inventor look to nature as a matter of course?* The ultimate goal is to naturalize nature-based innovation in the culture, to make biomimicry an everyday part of inventing. But my real goal and secret agenda is to increase respect for the natural world so we treat it and each other and ourselves as if life is a treasure—which it is.

Who do you naturally treat well, like they are a treasure? Obviously, it's your children and people you are here to care for, but beyond that it's people we respect. Our mentors. Ray was a mentor to so many because he was able to articulate his dream of a better world, and begin to put it in place so that we could see it, too. Mentors show us what is possible, and we're so grateful to them.

The process of biomimicry is realizing that the planet is full of incredible organisms that can and should be our mentors, and seeing them in that light leads to respect and to better behavior. Once we start to regard nature as a mentor, I have to believe we will fundamentally change how we treat all life. It all begins with respect.

The other table was looking at how some bird feathers interlock with hooks and barbules, so that when the bird ruffles its feathers, it can smooth them back down with its beak. The feathers are connected almost like a zipper on a jacket (somewhat similar to the innovation team's first design attempt, so they were not entirely off base). Janine recalled what happened next: "We are telling this story about bird feathers, and one person stood up and said, 'What if we attach carpet tiles to each other?' Then someone at the gravity table heard that and said, 'And gravity will hold the whole thing down because it will be attached as one.'"

The solution was a 2.5-inch by 2.5-inch adhesive square that Interface would patent under the name TacTiles. This square is stuck on the underside of carpet tile at every corner where tiles meet, linking them together. Because the carpet tiles weigh so much and their backing material has a high friction coefficient on the floor, gravity is able to hold the linked tiles in place as people and objects move across them. Importantly, the TacTiles do not off-gas VOCs, and a life cycle assessment showed that they have a 90 percent lower environmental impact than traditional glue—90 percent! Interface all but eliminated a category of emissions with one single breakthrough.

Interface has used biomimicry with great success in other ways. A now best-selling product line called Entropy was discovered after the design team considered how nature would design a floor covering. They observed that a forest floor has a clear design aesthetic, blending leaves, twigs, ferns, moss, and other components into a beautiful understory. That said, not one square foot of forest floor looks exactly like another. Instead of repeating patterns, nature invites randomness.

Interface's design team then created a carpet tile where no two cut tiles were exactly the same, but all fit within the same color scheme. The tiles could be randomly laid down since there was no pattern, decreasing installation time. If one tile became worn or stained, it could be replaced by any other without a noticeable sore-thumb effect. Perhaps most important, because randomness was designed into the carpet tiles, defects in manufacturing were not noticeable. Carpet tiles simply could not "misprint," so the off-quality of Entropy products is much lower. Waste goes down and profits go up.

As Interface has shown, biomimicry offers a transformative approach to innovation, showing that humans have been "doing it wrong" all along in quite a few ways. Nature's design elegance trumps human ingenuity in nearly every way. Just consider the design principles that Janine offered in her book:[2]

Nature runs on sunlight.
Nature uses only the energy it needs.
Nature fits form to function.
Nature recycles everything.
Nature rewards cooperation.
Nature banks on diversity.
Nature demands local expertise.
Nature curbs excesses from within.
Nature taps the power of limits.

If all innovation honored these principles, we might never have had unsustainable businesses in the first place. T (technology) would have been in the denominator the whole time.

Let us zoom in on one of these principles: Nature recycles everything. Back in chapter 1, Ray imagined the technologies of the future for Interface, technologies that "will enable us to feed our factories with recycled raw materials—closed-loop, recycled raw materials that come from harvesting billions of square yards of carpets and textiles that have already been made." Interface knew they could not wait for this technology to arise in its own right. They had to innovate it.

Interface's carpet tile has two main components: the backing and the face-fiber yarn. In order to recycle carpet, the two have to be separated, as they are made of different material. Interface does not manufacture yarn, so recyclability of that requires cooperation with yarn suppliers. I will get to that.

Interface does manufacture its backing, so they focused innovation efforts there to start. One option for recycling the backing was to melt it down into a liquid plastic, then repour it into sheets of recycled backing. There were two big problems with that, however. First, liquefying the plastic backing requires a massive amount of energy. Second, the polymer would begin to break down at those high temperatures, impairing the performance of the recycled backing.

It turns out both problems can be solved with less heat and more ingenuity. First, the backing is crumbled into small pellets, which are then evenly distributed on a conveyor belt. The conveyor belt passes through an oven that warms the pellets enough to make them soft. Then a roller compresses the soft pellets into a continuous sheet of recycled backing with performance characteristics as good as backing from virgin materials. The new backing method was called Cool Blue, and it rapidly increased Interface's closed-loop capabilities.

Innovation does not exist in some utopian bubble within a company, however. I spoke with Dan Hendrix about some of his difficult decisions as CEO of Interface, and he was quick to point to Cool Blue as one. "We were in violation of bank covenants at the time," Dan reflected. "Our innovation people brought Cool Blue to me saying it would cost $15 million, and I didn't think it would work and I knew it would take a long time."

The dot-com bubble had burst, challenging Interface's financial position with its creditors. Making an optional $15 million investment in something like this was about as financially unattractive as anything you could imagine. After some careful consideration, though, and not a little bit of heartburn, he ultimately said yes and funded the project. Ultimately, the company's sustainability culture oriented around long-term decision making demanded it. "David Hobbs [then a senior vice president at Interface] said to me, 'Are we going to do this or not, Dan? We keep telling people we are going down this road, and you created Mission Zero. We can't get there without making investments like this,'" said Dan. "There was no ROI, but Ray and sustainability had taught me to change the lens on the ROI. If you put it that way, I guess we have to do it."

Innovation had become unshackled from short-term financial justifications. Interface was pursuing a dual purpose now: doing well and doing good. In the case of Cool Blue, the environmental benefits justified the lack of clear financial benefits. Moreover, Cool Blue has been beneficial financially in the long run, simultaneously creating goodwill in the marketplace while making Interface less reliant on virgin oil. Simply put, sustainability can help catalyze innovation by shifting the business from short-term thinking to longer-term thinking.

What about that yarn that I mentioned? Well, in this case Interface has shown that a business can use its purchasing power to catalyze innovation in its supply chain. Remember, Interface does not manufacture yarn but rather buys it from suppliers. If the yarn was going to have recycled content, the suppliers were going to have to do the recycling. So Interface asked them.

"We had a meeting with DuPont, Universal, Aquafil, and some other yarn suppliers," Dan recalled. "We basically said, 'If you are not going to give us recycled content, you are not going to be our supplier.' At the time, DuPont was about 96 percent of our yarn purchases, and they said they could not find the business case for sustainability. Today they are 3 percent."

Other suppliers like Aquafil stepped up to the plate, and now Interface is able to shave off the face fiber from old carpet and return it for recycling. Closing the loop requires partnerships, and the yarn suppliers willing to be good partners were rewarded. Remember nature's design principles? Nature rewards cooperation.

Recycled content, along with bio-based content, does present some novel challenges for a manufacturing company like Interface. Once again, these challenges can be viewed as opportunities by innovators committed

A Conversation with Jay Gould

Jay Gould is the president and CEO of Interface, Inc. He joined the company in 2015 and became the third CEO in the company's history in early 2017.

JOHN: What was your career path before you joined Interface?

JAY: The first half of my career was focused on learning about global business, learning how to build organizations and brands that create value for multiple stakeholders.

I really had my own spear-in-the-chest moment in 2002 when I was the chief innovation officer of Coca-Cola and the CEO asked a small group of people to look at how we could spark magic back into the company—find the why behind the what. Legendary Coke CEO Roberto Goizueta used to say we exist to create shareowner value. That's not a stirring notion to get out of bed every day—there has to be a bigger why to that what.

We studied Interface, and though I never met Ray personally, he dedicated time to people who were on this exploration process with me and because of that, he had such a huge influence on the second half of my career. I was on this journey to help companies be purpose-driven, and we followed a lot of lessons Ray taught.

One of those lessons was, *You have to earn your right to pursue your sense of purpose.* Earning the right means you have a healthy business.

I was always fond of him saying that the next sale is the next heartbeat of the company—you have to keep the blood flowing through the body or whatever else you're doing positively can't happen. When did you get connected to Interface directly?

In late 2014 I met Dan Hendrix when he was looking for someone to help to reinvigorate the performance of the company and reenergize the people after Ray's passing in 2011.

Dan and I had planned to meet for a half hour or so and ended up spending two hours together and kind of fell in love with each other. Like Ray, he had dedicated his life to the development of Interface and he needed to know that I understood the values and distinctiveness of Interface.

Have there been any surprises—good or bad?

I was pleasantly surprised at the global consistency of the culture. When you go around the world, Interface feels like Interface.

The other surprise, looking back after three and a half years, is how Interface has changed me. I call myself "the accidental environmentalist" because I didn't join with a deep commitment to sustainability. I wanted to get into this company because I felt like it was a great platform for value creation; but it has opened my mind and heart in a way that had not been done before. It is humbling. It takes a hell of a lot of courage to run a company like Interface because you have to make commitments for which you're not prepared.

Have you found the sustainability mission to be authentic?

Clearly it is—a great many of our people join because of our commitment to sustainability and others have been caught up in the power of it, so it's clearly and deeply embedded. Everyone has a role in making a difference in our sustainability journey; whether you're working on the line in Thailand or selling carpet in Russia, we are all part of it. That is what makes it authentic.

What is your philosophy about why a corporate culture really matters for results?

I think culture and results are inextricably linked; they are both essential to run a purpose-driven business. I'm motivated by both but personally I'm probably more motivated by a healthy culture, which will ultimately deliver great results if you have a good business strategy. Being authentic means that you have to start from the inside and then take it outside—particularly when dealing with the increasingly transparent world that we live with [on] the internet.

Is corporate culture something that can be monetized and valued, and if so, any idea how valuable it's been for Interface?

I would say we've yet to harvest all the value that is there. *Monetize* is an odd word because it seems like it's a onetime transaction and I think the value of this is creating a truly sustainable business model that creates more and more value every year. When I talk about Interface now, I talk about wanting it to be the world's most valuable interior products and services company, and the only couple words I added to Ray's original words were *most valuable*. I don't want to be the largest but do want a dollar of our earnings to be worth more than a dollar of everyone else's. I also want our other chief stakeholders to think we're the most valuable. So it's not just about the financial markets, but do our customers think of us as the most valuable? Do our employees? Does Mother Nature?

to sustainability, rather than restrictions. Ultimately, it comes down to three component parts of industrial design: product design, process design, and chemistry.

When using virgin raw materials, it is fairly easy to build product and process design around the chemistry. As John Bradford explained, "Generally, you build machines around chemistries that have been the same for the last hundred years." In Interface's case that includes nylon face fiber. Nylon has been around since the 1930s. "This new journey is very different," he continued. "You are not working with pure chemistry anymore, but contaminated chemistry. Anytime you work in a recycled or bio-based area, the chemistry is never perfectly pure. You have to have a tolerance for imperfection, which is generally not taught in machine design. Ultimately, you need to design machines that give you enough degrees of freedom that the chemistry can move and the product design doesn't change too much."

In other words, new innovation was needed to make products that were just as good, but with recycled and bio-based components. The original motivation might have been the environmental benefits. At the end of the process, though, Interface has ended up with a more resilient and flexible manufacturing approach that can utilize a more diverse supply chain. Sustainable thinking can turn industrial design upside down, with some surprising and positive results. Again, innovation becomes much more about opportunity—or perhaps a better word is *abundance*.

"Somehow," John said, "we have this idea that innovation is about managing scarcity, because that is how we are attuned. It is really a function of abundance and choice, and the best innovators in the world have endless ideas of what we can choose from. Sustainability accentuates and leverages that further."

So why does Interface excel at innovative thinking? Ultimately, it is the product of three things, represented by Dan Hendrix, Janine Benyus, and John Bradford. First, you need big and bold leadership that encourages transformative innovation. Dan, right along with Ray, was committed to Mission Zero. He was willing to walk a path before he could see the way. As a result, he and Ray empowered Interface's research and development team to swing for the fences.

Second, you need new, outside-the-box thinking. The old form of technological innovation is what got us into our environmental mess in the first place! For Interface, Janine and biomimicry have been the most inspirational muses we could have imagined.

Finally, you need really smart and committed people. John and everyone who has worked with him know that they are doing more than just making a better carpet tile. They are creating a model for the rest of the industrial world to follow. They have bought in to a higher purpose, and that higher purpose has created an employee culture worth more than gold. The power of that culture is our next learned lesson.

Lesson Number Three: You Can't Get Far without a Cultural Revolution

Jim Hartzfeld's name has popped up a number of times in this book. He was the person who prodded Ray to give his first environmental speech back in 1994, and he helped organize and execute the global sales meeting in 1997. In the years that followed, Jim's influence at Interface only grew as he assumed more and more responsibility for sustainability efforts, including the role of vice president for sustainable strategy. He was at Ray's right hand every step of the way, and no one has as complete a perspective on the transformation that Interface underwent as Jim.

That is why I wanted to know from Jim what the culture was like at Interface before Ray had his epiphany. The answer, in a word, was "Bad." Jim joined the company late in 1993, and he talked about his first several months. "The level of fear and competitive infighting was astonishing. Within a few months I came to the conclusion that Ray may not necessarily have known much about what was going on. He had assembled a team of lieutenants under him, and I had evidence that some of them systematically insulated Ray from what was really happening. It was stunning."

I share that for the perspective that it gives—namely that Ray's epiphany may have come along at a time when Interface's culture was not thriving. As Ray indicated in his prologue to this book, the company had struggled from 1991 through 1993, and external leadership was brought in to stabilize the business. That external leadership stepped into a corporate culture that was troubled, and the years that followed were tense. Those years left the company hurting, many of its people squabbling, and Ray largely out of touch with his people.

Along comes Ray's epiphany, and rather than shock his company into alignment around a new higher purpose, he shocked most people the other way. *Environment? Sustainability? Ray must be crazy.*

Jim shared a story of one particular employee who had climbed the Interface ladder on the manufacturing side of the business: "After a couple of drinks at an Interface networking event, he started getting misty-eyed. Here was Ray, the guy who had given him an opportunity that he never imagined, helped put his kids through school, allowed him to afford a house up on the lake and give his kids anything that they wanted—his hero had finally fallen. It was like he wanted to walk across that conference center, that ballroom, and just hug Ray and help him off to the old folks' home."

Another story from Jim shows a different reaction from some people inside Interface—resistance: "I got called to come visit the senior leadership team at Interface Europe less than a year into our journey. I'm pumped up and excited to say, 'Let's talk about this sustainability thing. We need your insights.' I jump on a plane and go to our Shelf Mills plant in Europe, and as I enter the big conference room there, I've literally never seen a more antagonistic room setup. The tables were out and the chairs where shaped in a U, with one chair at the opening of the U. That was my chair," he recalled.

"The head of manufacturing says something like, 'I don't know what you Americans are thinking, but this is absolutely ridiculous. First, business has no role sticking its head up on this moral kind of social issue. Anybody that sticks their head up gets it chopped off, so it's just dumb. Second, who are you as Americans, as pigs of the planet, to tell us what to do on the environment?' There was open antagonism, and some people were thinking Ray had lost his mind."

The transition took years. While a few people like Jim "got it" immediately, most did not. Ray had to patiently beat the drum, which is why "Tomorrow's Child" and the global sales meeting in Hawaii meant so much to him. They were markers that the culture was shifting toward Interface's new sustainability mission.

Interestingly, the shift happened at different speeds depending on the company department. I asked Dan Hendrix about the change he saw in the manufacturing side compared with the sales side of the business. "The shop floor people got it first because of QUEST," said Dan. "They could actually see the changes that we were making at their recommendations. The sales force was the hardest to reach on sustainability. As CEO, I always talked about how you have to make this your own journey. Interface can't give you a shot in the arm about sustainability and you go show it to your customers. You've got to be inquisitive and eat and breathe sustainability if you're going to be authentic with your customers."

A culture is not a company's mission statement. It is not green building certifications slapped on the side of a company's facilities. You will not find culture in an employee manual or marketing pamphlet. You find culture in the hearts and minds of employees, manifested in how and why they do their jobs. Dan is right: Interface's people had to make sustainability their own journey.

Viewed in that light, Ray did not change Interface's culture. The people of Interface did that. Ray's job was changing one person's mind, and then another's and another's. Sustainability started with him, but it did not stay with him.

Daniel Esty and Andrew Winston, in their influential book *Green to Gold*, observed how important it is to move sustainability into employee culture if leadership wants to benefit from sustainability: "For fresh thinking to take hold and generate Eco-Advantage, the practice of looking at choices through an environmental lens has to be embedded throughout the organization. Relying *only* on a champion is a doomed strategy. Real success requires engagement from the top of the organization to the bottom."[3]

How has Interface benefited from an authentic sustainability culture? Well, to start, it helped put a stop to the infighting. Jim Hartzfeld told me of a conversation he had with Ray a couple of years after Ray's epiphany, when he still was not quite sure what would come of it. He said, "If nothing else, Jim, having a big purpose like this will help lift people's eyes above the petty, internal, competitive garbage that happens in every company." Maybe Ray was aware of some of the cultural problems his company faced after all?

More substantively, a culture engaged around a higher purpose aligns employees in multiple ways. For instance, a business benefits when its employees are willing to collaborate, both within individual teams and across teams. Strong collaboration results in more work being done in a shorter amount of time. An aligned culture promotes such strong and willing collaboration.

Similarly, an aligned culture can build common ground and shared values where they might not have existed. Consider Interface's sales and manufacturing associates. They do very different jobs in very different locales. Other than receiving a paycheck from the same company, those groups might not have much in common. Sustainability has been meaningful common ground for them at Interface.

This common ground can have real and practical benefits. Interface's sales associates are frequently sharing Interface's sustainability efforts with clients and prospective clients, and many of those people make the trip to west Georgia to tour Interface's manufacturing facilities. When they see that folks on the shop floor are living the same values that the sales associates have been preaching, clients come away impressed. The resulting goodwill is incredibly valuable.

Employees aligned around a common purpose are also much more willing to learn from, and share with, one another. When everyone across the company was tasked with reducing waste, a breakthrough in Australia would be copied in Europe and the United States to the greatest extent possible. Sustainability has been a *lingua franca* that has helped break up the silos that might otherwise have existed. To borrow Jay Gould's words: "Interface feels like Interface everywhere you go."

Sustainability has also played an important human resources role at Interface over the years. Every business out there would love to have a mechanism that attracts the best talent, and it turns out that a lot of brilliant people want to work for a company with a higher purpose. As Jim Hartzfeld put it, "Nobody would have known about Interface outside of a few architects and designers if it hadn't been for sustainability. We had multiple candidates applying for jobs with Ivy League, Stanford, and Berkeley caliber credentials, fighting to work to help sell carpet. People sought out Interface to get whatever job that they could, just to become a part of our purpose."

I believe that a company is only as good as its people. If I am right, then it means the better the talent a company attracts, the better it becomes. Further, sustainability has also helped retain the best talent at Interface. If a company's approach to talent acquisition is to simply pay more than its competitors, that company is always at risk of its employees being poached by a higher bidder. Employees who care about a company's purpose, on the other hand, might never be poached.

A higher purpose can also get the best out of people. Jeffrey Hollender, co-founder and former CEO of the environmentally minded cleaning products company Seventh Generation, observed this benefit at his own company. He wrote, "When a company's self-conception doesn't extend much beyond its financial objectives, it more than likely won't stretch people's ambition and drive."[4] With a higher purpose, a business can get the best out of its people.

So why has Interface been able to build and sustain a culture of engagement? Note that I did not say anything about employee passion. I have not used the word even once, though I commonly hear the idea discussed in the context of employee satisfaction. It seems as if everyone is being encouraged to go and find their passion.

I think a focus on passion is misguided. What if your passion does not create an employment opportunity? If everyone was following their passion, would we still have people willing to clean toilets, manufacture thumb tacks, or run our water treatment plants? It is not feasible for everyone to find their passion at work.

If passion at work is a luxury, though, purpose at work should be a prerequisite. Purpose has been the key to Interface's cultural success. A higher purpose can be universal, and it can be the thread that stitches together the disparate parts of a company. Interface has proven the internal benefits that come from a sustainability purpose. It has proven the external benefits as well.

Lesson Number Four: Goodwill Increases Appeal in the Marketplace

I grew up in the suburbs of Atlanta, Georgia. You can go ahead and translate that to, "I grew up in Coke country." Every sporting event, every restaurant, and every vending machine served Coca-Cola. If you wanted a Pepsi, you needed to move.

I distinctly remember being a teenager and bragging to my friends that I could taste the difference between Coke and Pepsi, and that I thought Coke was significantly better. In truth, I could taste the difference! Looking back on it, though, I cannot imagine a more inane point of pride. Even if they are different, that difference is minimal and trivial. There I was, though, disdainfully dismissing Pepsi as a product because of a hometown pride in the local brand.

Brand loyalty is tremendously important to any business. Often, that loyalty is earned through differentiation, whether perceived (as was the case with my teenage soft drink preference) or actual. When a customer believes that one product is better than another, they choose the preferred one.

Which leads to a tricky little question. How does a customer define the word *better*? Or asked more directly, on what grounds do businesses usually compete in their efforts to appeal to consumers?

Traditionally, it comes down to two factors: quality and price. When a customer picks up a product, they will decide if they like it and look at the price tag. Price is the easy part, in that less is more. Quality is a bit more complicated, consisting of factors ranging from aesthetics to durability to performance to convenience.

Occasionally, though, product differentiation goes beyond the realm of price and quality. When that happens, the playing field can shift in dramatic ways between competitors. A well-known example is Toms and its One for One pledge—one person in need is supported for every product purchased from Toms. All of a sudden positive social impact is a significant differentiating factor for the group of customers who care about that pledge. The differentiating factor can be negative as well. Think back to the backlash against BP following the *Deepwater Horizon* oil spill in 2010.

Jeffrey Hollender said it well in *The Responsibility Revolution*: "When organizations stand for something big—something that truly matters to people—they sharply differentiate themselves from their competitors. You can't make a difference if you're playing the same game."[5]

Ray changed the game in 1994. Sustainability may have started as a moral imperative for him, but it quickly became a business strategy. I think he and the company were surprised by just how positively the marketplace reacted. "The whole thing that got underestimated in all of this is the goodwill in the marketplace with our architecture and design firms. It set us way apart," Dan Hendrix shared. "About one-third of the architects and designers really cared. If it were a jump ball, we were going to get it and they were going to work with Interface no matter what."

To me, that is a stunning percentage. As Interface's sales associates know very well, architects and designers represent a massive portion of the floor covering market. If they like you, business is good. Brand loyalty among a third or even just a tenth of what's known as the A&D community represents millions of dollars of revenue.

Admittedly, Dan's estimation still suggests that most of the marketplace is not willing to give Interface business primarily for its sustainability commitments. From a practical standpoint, environmentally minded companies need to be strategic with their sustainability marketing, a sentiment that is explained well by Daniel Esty and Andrew Winston: "Marketing the green aspects of a product can be a tough proposition. Most successful green marketing starts with the traditional selling points—price, quality or performance—and only then mentions environmental attributes. Almost always, green should not be the first button to push."[6]

I wish sustainability were the primary motivating factor for every customer, but that is not yet realistic. True sustainability needs to be just as competitive on quality without becoming non-competitive on price. That is why all of the lessons in this chapter need to be thought of as an integrated whole. A business cannot be sustained on goodwill in the marketplace alone.

So how has Interface remained competitive? Even before Ray's epiphany, Interface prided itself on a high-quality product, and the sustainability journey has only enhanced that aspect of the business. Interface might miss out on a sale if price is the only factor, but its operational and material efficiency have allowed it to compete for customers who do not account for environmental and social concerns. In other words, sustainability has earned Interface plenty of business and hasn't really lost it any.

To this point I have been considering goodwill only in the product marketplace. There is another type of market to consider, though—the capital markets. Here, too, sustainability has become a positive differentiator, but that was not always the case.

In the early days of Ray's epiphany, sustainability was in fact a negative differentiator in capital markets. In the stock market in particular, plenty of investors thought Ray had lost his mind (much as a portion of Interface's employees had). He kept talking about sustainability in late 1994, and the company's stock price kept falling. The perception that a business must choose between profit and the environment was strong, and it was not long before Dan Hendrix and Charlie Eitel told Ray to stop talking to Wall Street.

Eventually, though, investors began to realize that Interface was not falling on an environmental sword. As the company performed well in the late 1990s and then successfully weathered the dot-com bubble, they even began to think positively of this environmental vision.

"I think along the way the company became cool to investors," Dan said. "They saw we were different than all the other carpet companies. I think the lesson we learned around the stock market was that they hated sustainability, then got neutral on it, and then saw you were different than anyone else. Eventually, it helped give us a better stock price."

Other companies have benefited from this perception in capital markets as well, but I think these benefits are still very much the exception and not the rule. Environmentally and socially responsible investing is an emerging practice. I am encouraged, however, by the trends. Every year, more investment managers utilize an ESG (environmental/social/governance) investment lens, more portfolios divest from fossil fuels, and more data suggest that investment returns are as high or higher when these filters are used.

If I had to bet, these trends will continue, and we will see similar ones for product consumers as well. Would anyone predict that society will begin to care less about social and environmental issues in the years to come? I doubt it.

So why did Interface ultimately win the goodwill of the markets? Sustainability was able to deliver exactly what the markets wanted. For customers who care about environmental issues, sustainability differentiated

Interface. For customers who care primarily about quality, sustainability helped Interface make better products. For capital markets that care about financial returns, sustainability made Interface more attractive.

What if, someday, most customers looked at environmental impact before the price tag? What if most investment managers were as well versed in carbon footprints as they are in price/earnings ratios? That day may not be imminent, but I believe it is coming. When it arrives, companies like Interface stand to benefit even more.

One Tiny Little Word

Look again at the title of chapter 3. Ray called it "Doing Well by Doing Good." I think the most important word in that title is the smallest one, wedged right in the middle.

It does not say "Doing Well or Doing Good." That is the old myth, now proven false. Financial gain does not only come at the expense of environmental degradation, and environmental well-being does not preclude economic well-being.

It does not say "Doing Well and Doing Good." That is what I would call sustainability lite, which is not sustainable at all. If a company uses a social or environmental side project to excuse its harmful-but-profitable business practices, it does not get a passing grade from me.

It says "Doing Well by Doing Good." Interface is laser-focused on transforming the industrial world by creating the prototypical company of the twenty-first century, a company that enriches society and the environment through its operations. By doing this work for the good of us all, Interface has proven that it can also do quite well for its employees and shareholders. In all respects, sustainability is a better way.

CHAPTER TEN

The Prototypical Economy of the Twenty-First Century

B y now I hope that you are convinced that the prototypical company of the twenty-first century just makes good sense. Resource efficiency, game-changing innovations, an engaged workforce, and a loyal marketplace are all rather persuasive. Yet those reasons do not account for perhaps the most compelling argument for creating such prototypical companies: They can help lead humanity away from the environmental abyss.

Obviously, we need more Interfaces, and on that front there is some good news. Quite a few companies are committed to pioneering truly sustainable enterprises, and they have benefited much as Interface has.

The problem is, I do not think we can achieve sustainability at a truly global level just by replicating what Interface and its peer enterprises—Patagonia and Unilever, for instance—are doing. All of business and industry exists and operates in a broader economic system, so we need change at a systems level for it to be sufficiently significant and expedient to avoid catastrophe. We do not know how much time is left on the clock, but we know it is ticking.

Put another way, the creation of the prototypical company of the twenty-first century is not enough. We must also, collectively, create the prototypical economy of the twenty-first century.

Our Current Economic System

Join me in a thought experiment. Let us compare the present day to the time of America's founding fathers, focusing in on two general metrics: human quality of life and the health of natural systems.

From a quality-of-life standpoint, the differences are dramatic. Life expectancy is significantly longer today, driven primarily by lower infant mortality rates. Even with a global population that has grown tenfold, a higher percentage of people have easy access to daily essentials like food and water. Technological developments have radically transformed humanity's capabilities in terms of mobility, communications, and education. If I had to guess, there aren't many of us, in the developed world at least, who would trade in the present day for life 250 years ago.

From the standpoint of natural systems, the differences are just as dramatic. In the present day we have lost millions of acres of forestland. Rates of biodiversity loss have spiked as human development destroys ecosystems. Agricultural lands are losing topsoil at alarming rates. Stocks in fisheries around the globe are at critical levels. Toxic chemicals are now dispersed around the world, most of which had not even been invented 250 years ago. While I cannot say conclusively, I doubt that even a single natural system at the global level is healthier today than a quarter millennium ago.

Human quality of life has been increasing while natural systems have been declining, and these two trends are not coincidental. To the contrary, our current economic system is structured to drive these trends. As Ray explored in chapter 2, we have created massive amounts of wealth (in other words, "human prosperity") in the last one-fortieth of a second of Earth-week, all at the expense of our finite planet.

Just take a look at the fuel that keeps an economy moving—its capital inputs. In their landmark work *Natural Capitalism*, Paul Hawken, Amory Lovins, and Hunter Lovins listed four capital types: human capital (labor, intellect, creativity, and social relationships), financial capital (monetary instruments), manufactured capital (infrastructure and tools), and natural capital (resources and ecosystem services).[1]

In general the more there is of these capital types, the more economic activity can ensue. We call this economic activity growth, and we measure it with gross domestic product (GDP). These capital types can also be thought of as limiting factors to an economy. If an economy (whether local or global) has insufficient capital reserves of even one type, economic growth will be constrained by that factor. Two exceptions to this would be a barter system that does not rely upon financial capital and a fully automated world that does not rely upon human capital (substituting artificial intelligence, a form of manufactured capital, for it). I think we can agree that neither of these exceptions applies to our global economy, at least at this time.

Two of these capital types are clearly derivatives of the others. Financial capital is in essence a proxy for the other types, in that it has no value independent of that for which it can be exchanged. Financial capital allows for the purchase of additional labor, infrastructure, and resources. At a global level, financial capital will only be constrained to the extent one of these other capital types is.

The other derivative capital type is manufactured capital, in that it is the result of the past combination of human capital and natural capital. Wrenches, assembly lines, and airplanes do not grow on trees. Humans use energy and raw materials to make them. At a global level, manufactured capital is also only limited to the extent that one of the other capital types is limited.

At a fundamental level, then, long-term economic activity is constrained by human capital and natural capital. While neither of these capital types is derived from the other, one is clearly dependent upon the other. Life on Earth thrived long before humans arrived, and it will presumably still thrive should we no longer exist. Nature does not rely on us, but we certainly rely on it.

Borrowing again from *Natural Capitalism*, here is a partial list of the ecosystem services provided by natural capital: production of oxygen, purification of air and water, decomposition of organic waste, protection against harmful cosmic radiation, regulation of the local and global climate, and production of grasslands, fertilizers, and food.[2] For all of us humans to exist, we need those services. For business and industry to exist, we need those services.

While human activity impacts the stock of natural capital (we can both plant trees and cut them down, for instance), natural capital is not dependent on human capital. Conversely, humanity's reliance on a healthy environment means that the stock of human capital is dependent upon the stock of natural capital. *As a result, at a systems level for our current global economy, the primary long-term constraint to economic activity is natural capital.* All of the other capital types are either derived from or dependent upon natural capital.

That reality does not bode well for our global economy or for us. The increases in quality of life these past 250 years have come from the steady increase in economic activity, largely within the context of a capitalist economic system. Increases in health care, food production, transportation capabilities, telecommunications, and any other sector are the result of the steady hum of our global economic engine. The stock of human capital is higher in that there are more of us and we have become more technologically advanced. The stock of financial capital is higher, as seen by the increased flow of monetary instruments in the global economy. The stock of manufactured capital is higher, witnessed by the tremendous build-out of infrastructure and manufactured goods. The stock of natural capital, however, is much lower, and it is the primary long-term limit to economic growth. Since the industrial revolution, we have been converting natural capital to these other capital types and calling it "growth." This trend simply cannot continue. From *Natural Capitalism* again:

> *Capitalism, as practiced, is a financially profitable, nonsustainable aberration in human development. What might be called 'industrial capitalism' does not fully conform to its own accounting principles. It liquidates its capital and calls it income.*[3]

Eventually, this relentless liquidation of natural capital will reach a limit. What happens then? Ray answered that question when he shared Paul Hawken's story of St. Matthew Island in chapter 2. Our economic system will overshoot its limits and collapse.

That last sentence is a controversial one, as seen by the debate sparked by Donella Meadows, Jørgen Randers, and Dennis Meadows in their 1972 work *The Limits to Growth*, later updated and republished in 2004.[4] They explored the systems dynamics around growth in general and modeled potential scenarios for the future—including one based on business as usual, which showed that our economic system would eventually outstrip a variety of limits and collapse.

Their work was quickly criticized from a variety of sectors. Conventional economists, politicians, and industrialists questioned the accuracy of their model, which was perceived as projecting an imminent doomsday scenario. Effectively, the argument was, *We haven't crashed yet, so you must be wrong!* That argument is like throwing a ball into the air and saying, *Gravity doesn't exist, because the ball is going up!* It is a matter of timing.

Others acknowledged the basic logic of their argument but dismissed it as a threat too distant in the future to cause worry today. To that group of skeptics—and those that have followed in their wake, despite the very close trajectories of the business-as-usual scenario and reality over the last four decades—I would say, think again about our world 250 years ago. In the United States we fondly remember those who fought to create our country, enshrining values in our governmental system that persist today. We remain grateful to them long after they passed away. Will generations 250 years from now remember us so fondly? I fear they will look upon our economic system driven by excessive depletion of natural systems and say, "Shame on you."

Still others claimed that our economic system is not actually bound by the limits they identified. Running out of oil? We have natural gas! Running out of topsoil? We can use hydroponics! Everything has a substitute, and market forces will protect the system from encountering a limit. Essentially, scarcer and scarcer materials become more and more expensive, incentivizing a shift to a more abundant and less expensive substitute. When the substitute does not exist, it will eventually become lucrative enough to spur invention of the substitute.

This argument is a much more nuanced critique of the theory in *The Limits to Growth*, and it is very appealing to the foot draggers that Ray discussed previously. To understand why, let us consider the current transition in global energy supply. Energy from fossil fuels is steadily being replaced by renewable energy. As fossil fuel reserves are depleted, they become more expensive, and will only continue to do so. Renewable energy becomes cost-competitive, allowing it to scale and enjoy price reductions from economies of scale in the process. Eventually, renewable energy will entirely replace fossil fuels, and since renewable energy is not limited like fossil fuels, it will not be a constraint on economic activity. Ta-da! Our current economic system has an elegant self-correcting mechanism that steers it clear of any limits, or so the argument goes.

I want to be persuaded by this logic. I want to believe that price signals in the market will incentivize every technological development that is needed to save us from every natural limit. I want to be a foot dragger.

Alas, I am not, for two reasons. First, this self-correction only works in capitalism when the market reflects true costs. If we are not pricing a limit, the market will not account for it. Fossil fuels are priced according to the limit of their quantity, so capitalism will make sure we never run out of fossil fuels. What about global warming, though? Just like the price of a pack of cigarettes does not account for its true cost in terms of health consequences, the price of fossil fuels does not account for their contribution to global warming. Therefore, the market is not equipped to steer us clear of those limits. Neither are we pricing other ecosystem services, including oxygen production and water purification. Even if we could all agree to start pricing such ecosystem services, I am not sure that we could price them accurately enough to ensure sufficient self-correction. So much uncertainty exists around all of the ways in which human activity is harming natural systems.

Second, this argument presupposes that humanity will always be able to invent a substitute before a limit is reached. I will acknowledge that humanity might be that good at avoiding catastrophe. To me though, putting all of your eggs in that basket is mighty risky. What if topsoil loss eventually results in a sharp decline in global food production? Maybe we create sufficient synthetic agricultural techniques to fill the gap. Then again, maybe we do not, and humanity goes through the most extreme period of mass starvation it has ever witnessed. Or perhaps that synthetic agriculture unleashes a host of unintended consequences for our health or environment. I would strongly suggest we do all that we can to protect our topsoil, just in case.

Ultimately, I am persuaded by the logic of *The Limits to Growth*. I will not predict when limits will be reached or when our economic system will suffer the ensuing crash. I simply believe that we are going down a road that eventually takes us to that crash.

The way I see it, our economic system has three potential futures, and "status quo" is not one of them. Our economic system is fundamentally out of equilibrium, and as with all systems, it will eventually find an equilibrium point. A crash is one potential future, and the ensuing equilibrium point will not be pleasant in terms of human well-being. In that case, economic growth will take humanity past certain natural limits, and the resulting crash will drive humanity's quality of life well below current levels. How bad would that be? I do not know, though I am confident it will be significantly worse than any economic crash the modern world has experienced to this point. The system would not rebound this time, since excessive growth will cause overshoot and collapse. Systems do not return to their past growth trajectories after overshoot and collapse, so maybe it would be better for us not to find out just how bad it would be?

The other potential futures involve taking the off-ramp from our current road. Perhaps a better analogy is that they represent an evolution of our capitalist economic system. Both would, in theory, allow our system to settle into a sustainable equilibrium, and I would happily sign up for either future in a heartbeat. They may not be equally achievable, however, or have the same margin for error as we make the collective transition. For my money, I would bet on the second one being the better prototype, but I invite you to place your own bet.

A Conversation with Andrew Winston

Andrew Winston is a globally recognized expert on how companies can navigate and profit from humanity's biggest challenges, and the author of Green to Gold *and* The Big Pivot.

JOHN: Where did your environmental ethos come from?

ANDREW: I didn't grow up in a particularly eco-house or doing a lot of nature stuff, such as camping, or taking politically minded actions. But I had my own personal ethos about wanting to live a responsible life, and my career change came as part of the opportunity of finding myself unemployed and getting to this moment where I had to pick a new direction for my career. In the beginning I didn't have the language for it, but sustainability felt logical and responsible. I started digging into sustainability to get some underpinnings as to why, and it confirmed what was sitting in the back of my mind for a while: We couldn't continue the way we were going.

I recently found my old notebooks from when I was going through that change and my first information interview was with someone doing green consulting, and he says, "You should get to know this thing called 'sustainability' and here are a couple of books." And he suggested *Mid-Course Correction*. Ray's book was actually the first one I read and it was the beginning for me.

I knew that everything in *Mid-Course Correction* and *The Ecology of Commerce* was true—I couldn't un-know it anymore and that was my ethos. We don't know the exact timing of the systemic failures and challenges that will come. The planet is undeniably limited in certain ways. We are incredibly innovative in how we use resources on our endeavors. If you spend any time in business, you'll get a sense of the scale. We could literally use up everything. We could do it. Ray's book and Paul's book were my spear in the chest—clearly this is what business needs to look like and we're not going to thrive and survive this experiment in humanity and capitalism without figuring out a different way.

What's your take on the broader movement of sustainability in business?

I think Ray and Paul and the Lovinses started the battle, and twenty-plus years later we've won, to a degree. The largest companies in the world all have sustainability goals and all put out a corporate social responsibility report. There is no CEO—not even Exxon's—who doesn't know that he or she needs to talk about this. The problem we have is that the horizon we're shooting for—the stable climate and thriving planet—is moving away faster than the progress we're making in business. There's an article every damn week saying some change in the planet is moving faster than we thought. The Antarctic is melting faster and faster, so we're actually not winning. It's a very weird thing because both are true: We won the battle to get this on the corporate agenda, but the number of companies that have made it truly core to their business the way Interface and Unilever have is still tiny. The system of neo-liberal capitalism does not allow us to move really big companies in the way that Interface and Unilever have tried. It's hard to grapple with.

Are you optimistic?

How can you not be optimistic about the price of renewables and just the movement in the clean economy? It's truly remarkable. And I probably would have been at your level of optimism in mid- to late 2016 but we've seen a rise—and not just in America—but a rise of authoritarianism that actually threatens the world—and it does connect to this conversation because how we live on this planet includes equity and respect and freedom, and we are in the battle for our lives here. I spend increasing amounts of my time not in my chosen profession of corporate sustainability, but just getting involved to try and help us save our democracy. I am optimistic about the general direction where corporations are going, of course, but at the same time, even a lot of our favorite leaders are hardly always on the right side of everything.

The problem of our shareholder maximization system is pretty profound.

What do you think an evolution of capitalism would look like?

I've spent a lot of time over the last couple of years with Hunter Lovins and others moving into this realm of talking about the underlining economic narrative. There is a new kind of movement called the Wellbeing Alliance that is talking about new forms of capitalism and different economic narratives to fight the neo-liberal story. In that story, the dominant narrative is freedom above all and freedom of markets above all, and the only measure that matters for countries is GDP and the only measure for corporation is profit.

Instead, the purpose of a society and economy and of a government should be to maximize well-being. Part of that is economic well-being, of course, but it's hardly the only thing. We have to change that underlying story or we are only going to go so far so fast.

How receptive are people to your message these days?

There are a handful of us on the speaker circuit who are most often speaking to the non-converted. I'm not going to speak at Interface or Patagonia; I'm speaking to big, regular companies, and I'm still standing in front of ambivalent, and sometimes hostile, audiences. Almost every chief sustainability officer I know has dealt with some important executives who remained unconvinced, and there are key places in the organization where they know an executive can stop their agenda from moving forward.

The weird thing about being involved in corporate sustainability is that you live in this Venn diagram of barely touching circles where to the true green people, I am a sell-out corporate guy, and to the typical CFO in business I am a radical hippie. The overlap of the Venn diagram is getting a little bigger, but it's still a tiny percentage. Where you truly don't see any real long-term or major conflicts between sustainability and profit is when you encounter a CEO or business that actually sees sustainability as the driver of profit—that's part of what made Paul Polman so different. The way he talks about this. And that fluency that he and Ray both have had is still pretty unusual.

If someone is at a company that hasn't yet set sustainability goals, what are the starting points?

The way a lot of companies approach this is from a competitive mind-set—that is, what is our competitor doing and how do we get one step ahead? But it really shouldn't be like that with sustainability—there are actual biophysical thresholds that require the outside-in approach of science-based goals.

Sustainability needs to become the moral threshold that safety has become—it's about zero fatalities, not about outpacing competitors.

I typically start with what's coming (for a company): Here are the next megatrends, and what do they mean for you? Next is understanding your value chain: Where do you fit into this and where are the risks? What are the opportunities and threats upstream and downstream? That systems view is a really critical part of that first analysis, and then you get into organizational readiness and understanding where people are coming from inside, and listening to your employees—especially the younger ones.

You come away with an approach that acknowledges the megatrends and the problems we need to solve in the world, and you understand how your company matches up against that: Are you enabling solutions that build a thriving world or not?

A good example is BASF. Over the last five years, they analyzed every one of their sixty thousand products and rated them on whether they are a part of the solution or not. They found a tiny percentage that were potential real problems for them and made a plan to either change what is risky about them or to exit those businesses.

And then there are a bunch that are kind of neutral and then there is a quarter of them they think could be "accelerants"—products the world needs now—and they are putting a majority of their R&D into that group of accelerators.

**When you talk about megatrends, is one of them
global warming?**

Yes. Companies that see the opportunity the easiest are those in the
clean economy space or in the value chain for the clean economy.
Those are huge growing industries.

What leaders are lagging on is some of the understanding of
what the extreme weather is doing to the world. I think the risk
side gets a lot of attention, but I'm not sure that many companies
and people have really thought it through. The big ag companies
are seeing their growing regions shift, but I think there's a lot of
denial about the non-linear effects of what could be happening
here and the sea level rise and increased storms.

When I help an executive who doesn't realize they could
serve a carbon-reducing world in some way with their business,
it is usually a lightbulb moment. It's happening the most in the
building and transportation sectors, and of course in energy. The
vast majority of energy being put on the grid globally is clean. If
you, for example, only make parts for a coal plant, your business is
going to disappear.

Decoupling Growth

As previously discussed, natural capital has been the primary fuel driving
economic growth since the industrial revolution, and it is also the primary
limit on long-term economic growth. A good analogy is a horse and wagon.
Our global economic wagon has been hitched to an environmental horse
that has been run hard. As that horse shows signs of fading, it is reasonable
to ask, *Can we hitch our wagon to another horse and give the environmental
one a rest?* That is the idea behind decoupling.

Decoupling growth from the environment would mean that the natural capital used in economic activity would only be that which is rapidly
renewable or already in use in our economy or waste streams. Such a shift

would involve massive changes in technology and business practices, and it is exactly what Ray was talking about with his rewritten Ehrlich environmental impact equation at the end of chapter 1:

$$I = \frac{P \times A}{T_2}$$

In other words, environmental impact equals population times affluence over T_2, which represents sustainable, and eventually restorative, technological practices scaled across the globe. All energy will come from renewable sources, with perfect cycling of materials to ensure that the stock of natural capital remains steady or increasing. In essence, it involves all of society successfully scaling Mount Sustainability.

"Wait a second," you might say. "Isn't that just what the foot draggers are talking about? Isn't that the point at which technological advancement saves us from environmental ruin?"

My answer to this is, "Almost." The end result looks very similar, but the means to reaching that end would be different. Rather than reliance upon business-as-usual and price signals in the markets to catalyze the change from unsustainable practices to sustainable ones, the change would stem from other forces, likely in combination.

For instance, changes in consumer behavior might be one such force. Imagine if every consumer considered environmental and social impact, both positive and negative, in every purchasing decision. If that happened, individual businesses that hoped to grow would have to compete just as hard on environmental and social performance as they do on quality and price.

Another force might be a massive cultural shift in business and industry itself. What if every business leader thought like Ray Anderson? What if every business leveraged its supply chain, only purchasing from suppliers with strong environmental and social performance, exactly like the customers I just mentioned? What if "bad actors" were shunned and shamed by the collective business community? Imagine that! I would love to see a world in which only the best corporate citizens were growing.

Another force might be regulatory change. Governments set the rules in which businesses operate. If those rules were structured to encourage businesses to compete on environmental and social performance, the pursuit of T_2 practices would accelerate dramatically.

Now imagine if all of these happened simultaneously? Capitalism will indeed have evolved, fully recognizing that our economic activity has multiple stakeholders, not just equity shareholders.

"Capitalism has gone astray by focusing on only one stakeholder, the investor," explained Jay Gould. In the industrial re-revolution, he asserted, "We need to move capitalism to a multi-stakeholder model. In Interface's case, our intention is to create value for our customers, employees, and shareowners, but also for the environment. I believe that corporate reform and governance reform is absolutely required to make this a reality."

Notice that Gould did not say *share* or *divide* value among the stakeholders. He said "create" that value. Far too often, people assume that a triple-bottom-line approach to business means that financial profitability must be capped, with that form of value diverted to environmental and social causes. Effectively, the financial profitability of companies subsidizes the other two legs of the stool. That is a weak form of the triple bottom line. The stronger, more robust form of it is this future vision of the economy, where strong social and environmental performance is an accelerant to financial growth. Rather than "some of each," it is about "more of all." This vision of the future is exactly what Ray Anderson had in mind when he wrote the first edition of this book.

I remain troubled, however. The idea of decoupling growth from environmental degradation leaves me wondering to what else growth might be recoupled. It will not be renewable forms of natural capital. While sunlight is essentially unlimited, other resource types are not. There is only so much timber we can sustainably harvest each year and only so many fish we can sustainably catch. Natural capital will always have some fundamental material limits. Other than the occasional meteorite, Earth is not getting any more "stuff."

We cannot hitch our economic wagon to financial capital or manufactured capital, either. Remember, these capital types are derivatives of human capital and natural capital. Their limits are a function of these other forms of capital. That leaves human capital as our only hope for long-term sustained economic growth.

How does human capital increase? Quite a few ways, actually. The obvious one is more people on the planet. Another is an increase in productivity, perhaps through lower unemployment rates or the average person working longer hours. Unfortunately, we cannot look to these forms of human capital for sustained economic growth. There are only so many hours in the day, and eventually human population will plateau, whether that is in the next thirty years or three hundred.

Human capital can also increase as our creative capacities increase. The more educated and innovative our species becomes, the higher our stock of human capital goes. That human capital unlocks more valuable forms of manufactured capital, which then positively reinforces economic growth as well as the stock of human capital. For example, the invention of the computer fueled economic growth at the time, while also increasing humanity's collective intelligence, creativity, and scientific discovery capabilities. The computer then paved the way for additional technological improvements like the internet, which also fueled economic growth and increased the stock of our collective human capital.

Perhaps our capacity for creativity and technological advancement could be the unflagging horse that carries our economic growth wagon off into an endless sunset. Who knows, maybe someday our innovative capabilities will allow us to settle other planets and shake off our natural capital limits here on Earth. Barring such a breakthrough, however, sustained economic growth will depend entirely upon our species getting smarter, more creative, and more innovative.

Then more innovative. Then even more innovative. On and on, again and again. What happens if the well dries up and we reach a maximum creative capacity as a species? Well, then our economic wagon will grind to a halt. The game will be up. Growth will stop, and this future economic system will no longer be able to deliver exactly what it is supposed to deliver: ever-increasing growth.

We are left with a prototypical economy of the twenty-first century (and beyond) that has a potentially fatal weakness, namely its dependence upon growth. I want to be very clear—this weakness might be overcome! Capitalism might successfully evolve toward a sustainable equilibrium and maintain its primary characteristic—ever-increasing growth with ever-increasing human well-being as a result. If this is the future in which you believe, I will not argue with you. We can happily work together to decouple growth from the environment.

At the same time, I think a cautionary approach to our future would require us to work on a Plan B as well. Perhaps the prototypical economy of the twenty-first century can be created without a potentially fatal weakness.

Perhaps it might be even better than Plan A. Which leads us to the other possible future for a sustainable economy—an economy that is not dependent upon growth at all.

Rethinking Growth

We should revisit the Ehrlich environmental impact equation and focus in on another variable, the capital A. It represents "affluence," which we can think of as average rates of consumption of goods and services. A common metric for the concept of affluence is GDP per person. Thought of that way, the environmental impact equation would read as follows:

$$I = P \times \left(\tfrac{GDP}{P}\right) \times T$$

Mathematically, the two P's in that equation would cancel out, resulting in an alternative equation stating that environmental impact is driven by economic growth and the harmfulness of the technologies that generate that growth:

$$I = GDP \times T$$

Reframed in terms of Ray's vision for sustainable enterprise, T_2 in the denominator, we are left with the following equation:

$$I = \frac{GDP}{T_2}$$

Here is how I read the essence of that equation—which, as a reminder, is intended to illustrate a concept rather than calculate numbers. Once business and industry have fully embraced sustainability, the only negative impact our economy could have on the environment would be economic growth itself. If we could decouple our economy not just from the environment, but from the need for growth, I believe we could return to living in harmony with our natural systems much sooner.

To my knowledge Ray never rewrote the environmental impact equation as I just have. I do know that he was thinking along the same lines, however. In February 2009, two and a half years before his passing, Ray rewrote the equation one last time, presenting it during a TED Talk in Long Beach, California. Here is how Ray further revised the Ehrlich equation:

$$I = \frac{P \times a}{T_2 \times H}$$

As he explained, the original *A* was capitalized, suggesting that consumption is an end in and of itself. Ray made the *a* lowercase, suggesting that it is a means to an end, and that desired end is really human happiness, represented by the *H* and understood as the satisfaction of basic human needs for all of us. "More happiness with less stuff," Ray said. "You know, that would reframe civilization itself, and our whole system of economics."

Others have advocated for similar changes. In *The Big Pivot*, Andrew Winston wrote, "We'll need to redefine what success looks like in our economy and particularly for large companies. Growth as a goal in and of itself needs to give way to *prosperity*, a much healthier, more mathematically feasible idea."[5] In *Doughnut Economics*, Kate Raworth (whose work we will return to shortly) wrote, "Western cultures seeking to oust the . . . goal of GDP growth cannot simply put an Andean or Maori worldview in its place but must find new words and pictures to articulate an equivalent vision. What might the words for that new vision be? A first suggestion: *human prosperity in a flourishing web of life*."[6]

These ideas can be perceived as threats to the current system, and plenty of people have strong beliefs in the virtues of that system. Therefore, I feel compelled to clarify what I am *not* saying.

I am not arguing that capitalistic, market-based economies are evil, and neither am I arguing that business and industry are evil. Far from! I give full credit to our current economic system for doing more to raise the quality of life than any other system. I also give full credit to our current economic system for being incredibly efficient. What I ask is that we acknowledge the weaknesses of our current system and imagine how it might improve.

I also am not arguing for that system that inspires fear in many Americans: communism, which can wrongly be thought of as the only alternative to capitalistic, free-market economies. I am simply saying that capitalism is capable of being made better. I am calling for an economic evolution, not a revolution.

Finally, I am not saying that it will be easy to create, or even imagine, what this future economic system will be like. In fact, my words might have left many of you feeling unsettled right now, asking, "Should I not want my stock portfolio to go up? How am I supposed to plan for my future? Real people suffer when the economy is not growing, so what about them? And are businesses supposed to just stop competing in the marketplace? How can any of this work?"

Those are all fair questions. The world has never seen a global economic system oriented around human prosperity, so we are talking about inventing something new. Science-fiction writers from fifty years ago would have struggled to imagine a future in which people commonly carried supercomputers in their pockets as we do today, but that merely reflected a failure of their imaginations. Just because it is hard to imagine does not mean that it cannot happen.

Also, "how this all works" does not necessarily need to be determined in advance. I am talking about changing a goal, and the strategies and tactics that are best at accomplishing that goal can be discovered in time.

Consider an analogy to the game of chess. It is an ancient game, with the modern rules having been established for more than a hundred years. Even if you have never played it, you likely know the goal of the game. The first player to capture the opponent's king wins. Expert chess players have developed intricate strategies and tactics in pursuit of that stated goal.

Now imagine that we changed the game in one simple way. Instead of winning by capturing the opponent's king, the winner is the first to capture both of the opponent's knights. I am not changing how the pieces move. I am not changing the size of the chessboard. I am not changing the competitive drive that players have to win the game. I am just changing the definition of success. Entirely new strategies and tactics would need to be developed for this new goal, and even the best chess players in the world would need to tinker and experiment before discovering the best new strategies for "Chess 2.0." Without knowing in advance how it would all work, can we similarly evolve our global economy into a 2.0 and maintain the structural components that have made our current system great-but-not-perfect? I hope so, and I think so.

Even still, some brilliant people are imagining how this future system might work, and Kate Raworth is one of my favorites. She conceives of a new form of economics that she calls doughnut economics, a reference to

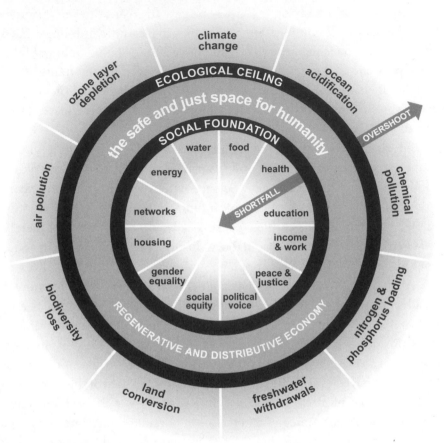

Kate Raworth's Doughnut.
Courtesy of Kate Raworth.

the visual depiction of this system being two concentric circles with a hole in the middle, looking like a typical American doughnut.

The inner circle represents the various forms of human needs, from access to food and water to civil rights such as political voice. Ideally, all of humanity has its needs met by this future economic system; no one should fall out of the inner circle into the doughnut's hole.

The outer circle represents environmental limits, like those conceived of in *The Limits to Growth*, including air pollution, ocean acidification, and land conversion. This future economic system would need to not exceed these limits, depicted visually by being within the outer circle.[7]

Layered together, these circles create the doughnut within which humanity and its economy need to stay. Success of this economic system would be getting within the doughnut's inner and outer boundaries. Long-term success of the system would be staying there. Here is how Raworth described the transition to this new goal: "At this point in human history, the movement that best describes the progress we need is *coming into dynamic balance* by moving into the Doughnut's safe and just space, eliminating both its shortfall and overshoot at the same time."[8]

Of note, Raworth's doughnut is not predicated on equality but rather on equity. Success means that everyone's needs will be met. There is no problem with people enjoying luxuries that meet their own tastes, so long as their consumption of goods and services does not take the system beyond environmental limits. This new system is not intended to deprive people but to empower them.

In any economic system, the actors within it are people, just like you and I. The system operates with basic assumptions about who we are as economic actors and how and why we engage in the system. For instance, our current system assumes that people are self-interested and rational economic participants. These assumptions have worked reasonably well, and humanity has dutifully played the part to an extent. To me, though, and to Kate Raworth, the current economic assumptions do not quite capture who we really are as human beings.

Raworth offered up five broad shifts in these assumptions under this new economic paradigm: "First, rather than narrowly self-interested, we are social and reciprocating. Second, in place of fixed preferences, we have fluid values. Third, instead of isolated, we are interdependent. Fourth, rather than calculate, we usually approximate. And fifth, far from having dominion over nature, we are deeply embedded in the web of life."[9]

Now, that is a better understanding of who we are as a species! So why not design an economic system that incorporates this understanding of economic actors? The system can involve growth, but it would not be dependent upon it. As Raworth said, the system would be growth-agnostic: "By agnostic, I do not mean simply not caring whether GDP growth is coming or not, nor do I mean refusing to measure whether it is happening or not. I mean agnostic in the sense of designing an economy that promotes human prosperity whether GDP is going up, down, or holding steady."[10]

Are you still skeptical? I understand if so. Once again, such a system has never been created before, at least by humanity. That does not mean that such a system does not exist, however. Ray was fond of quoting his friend Amory Lovins: "If something exists, it must be possible." So what is our model?

We need look no further than Mother Nature. What is an ecosystem if not an economy? Species after species, life-form after life-form, all exchanging and utilizing valuable materials each to its own needs. Does an ecosystem experience growth? Sure it does! It also contracts as needed, prioritizing the well-being of the entire system over endless growth. When it comes down to it, maybe creating the prototypical economy of the twenty-first century is not about saving the environment. It might just be about learning from it. Somewhere out there, Janine Benyus is smiling.

The Inevitability of Change

If I am right, and I think I am, we have three possible economic futures ahead of us: a crash, sustainability with endless growth potential, or sustainability that becomes growth-agnostic. As long as we avoid the first one, I will be happy. Doing so will not involve a sudden change, however. We cannot all collectively decide to redesign our economic system over-night. It will truly be an evolution if we are to avoid collapse.

As in biological evolution, the system will respond and adjust to certain stimuli. If we are smart, we will look far enough ahead to see the economic stimuli that are coming and develop strategies to use those inflection points positively. Doing so will ensure our economic system evolves toward sustainability of some sort and not toward the crash.

I do not have a crystal ball, but I do have a good idea of some stimuli on the horizon. One is intergenerational wealth transfers from baby boomers to their children. Another related stimulus is the mind-set of millennials who are about to move into leadership positions in every sector.

As artificial intelligence continues to develop, it, too, will have tremendous implications—for manufacturing capabilities, transportation, advanced computing, and labor markets. Speaking of labor markets, what will happen when population growth does stall? Projections have that happening as a result of declining birth rates within the century. Our global economy has never encountered a flat or declining population rate, which would certainly be an important stimulus for change.

In the next two chapters, though, I want to focus on the two stimuli to our economy that I believe will prove defining in the twenty-first century. If we are successful in crafting the prototypical economy of the twenty-first century, it will be because we responded brilliantly to these coming challenges. We will not have just "dodged economic bullets," but rather we will have seized upon these challenges as economic opportunities. The first challenge is resource scarcity, and the corresponding opportunity is called the circular economy. The second challenge is global warming, and the corresponding opportunity is called drawdown. In both cases, I believe that business and industry will take the lead in embracing these opportunities.

CHAPTER ELEVEN

Bending the Line

A single trip is all it took for London to become my favorite international city. There is an incredible combination of history, greenspace,
and diversity, and it is easy to get around. It didn't hurt that I had a martini
at Duke's Bar, which was a regular hangout of Ian Fleming, the creator of
the legendary fictional spy James Bond. It is rumored that Fleming gave
Bond his love for martinis because of that bar, which I believe considering
that they make the perfect martini.

I was in London to work, though, not play. The Ellen MacArthur
Foundation was hosting their annual summit, a conference exploring the
circular economy, why we need it, and how we are getting there. They
were holding the event at the Roundhouse, a performing arts venue
in Camden Town that has been host to the Beatles, Led Zeppelin, the
Rolling Stones, Jimi Hendrix, and plenty of other noteworthy perform-
ers. I have to admit to being slightly awed by that history. The building
has even deeper roots, having been constructed in 1847 as a facility for
maintaining the London and North Western Railway's locomotives. For
a conference that focuses on keeping materials in productive use as long
as possible, it seemed fitting that we were in a building with such a long
and varied history.

I walked into the event space not quite sure what to expect. The
Roundhouse's interior retained the industrial feel from its origin. Up to
that point, I had intellectually understood how exciting the transition to
the circular economy is. Now for the first time, I got to feel how exciting
it is. The Ellen MacArthur Foundation had transformed it into a modern
space with clean trappings and a hip vibe. Pulsating music filled the large,
circular room that they had bathed in soft red and blue light. Rather than
use a single projector screen, they had two large screens that joined at
an angle, so that both sides of the venue could clearly see. Visually, the
screens looked like two faces of a massive cube set immediately behind the

presentation stage, as if the foundation were saying, *We are here to literally think outside the box.* And of course, we were.

A Bit of History

Way back in chapter 1, Ray discussed how the technological practices of the first industrial revolution are linear, based on a take-make-waste model. When describing technologies of the future, Ray used the term *cyclical* rather than *circular*, but they are in substance the same. I have chosen the latter terminology as it has been broadly adopted by the sustainability community.

Regardless of the label, circular economy ideas predate the thought leadership of both the Ellen MacArthur Foundation and Ray Anderson. Some of the earliest credit belongs to Walter Stahel and Geneviève Reday-Mulvey. In 1976 they submitted a report to the European Commission titled *The Potential for Substituting Manpower for Energy*. During their research, they realized that more energy was used in the extraction and production of industrial materials (such as steel) than the energy used to assemble those materials into durable goods (like machines and buildings). With labor, the inverse is true—more labor goes into assembling materials together than simply extracting them from the Earth. They concluded, therefore, that energy could be saved and jobs could be increased if remanufacturing and reconditioning activities were prioritized over new manufacturing, since the former does not require the production of new industrial materials. In other words, repairing used durable goods saves energy and creates jobs.

Stahel continued to build upon these initial insights, and in 1981 he wrote a paper titled "The Product-Life Factor." In it he visually depicted replenishing loops of materials that keep those materials from becoming waste. He called for an economy "based on a spiral-loop system that minimizes matter, energy-flow and environmental deterioration without restricting economic growth or social and technical progress."[1] Arguably, he drew the first circle in the circular economy movement, and his lifetime of work in the space has made him one of the "godfathers" of sustainability.

By the time Ray read *The Ecology of Commerce*, others had advanced the field further, including Bill McDonough, Michael Braungart, Gunter Pauli, and David Pearce. As Ray shared, McDonough was the one who introduced him to the concept of "cradle-to-cradle," which had become a part of the circular economy lexicon by 1994. Their work inspired more and more people to open their eyes to the structural flaws of our linear economy. As a result, Ray was jumping into a river that was already moving, and picking up speed.

Ray's contribution to the movement was twofold. First, he validated the theories of all those who had come before. Up until then it was merely theoretical that a company could pursue a circular framework and succeed. While Interface is not a perfectly circular company (yet), it is clear at this point that material efficiency, energy efficiency, sustainable design, and end-of-life recovery technologies for Interface's products have been beneficial to the company.

Ray's other contribution was to visualize a circular company, rather than a circular economy. Most circular theories at that point utilized a macroeconomic lens, not a microeconomic lens. Look again at The Prototypical Company of the 21st Century as visualized by Ray, in chapter 5 (see page 87). All of those loops are in the framework of a single company and its social and environmental impacts.

While Ray and Interface were busy constructing this prototype in the mid-2000s, Ellen MacArthur was busy breaking records. Throughout that decade, MacArthur competed in solo long-distance sailing competitions. Perhaps her crowning achievement was a circumnavigation of the globe in seventy-one days, fourteen hours, and eighteen minutes, setting the world record at the time.

"I never thought I'd do anything other than sailing," she told me. "I never had a goal or aspiration to do anything other than sailing. Sailing was my entire life for many, many years."

During her voyages, though, MacArthur began to ask herself some questions. I suppose a lot of questions might pop into a person's mind when alone at sea for months at a time. In her case, the dominant one was, "Is the finite nature of resources that you have on a boat any different than the finite resources that we have available to our global economy?"

She understood very well how finite the resources were on her boat. When she would begin a voyage, she essentially possessed all that she could have. There would be no resupplying until the voyage ended. In her words, "You manage what you have, and you watch the supplies go down every day."

She saw our global economy as functioning the same way, but without the possibility of a resupply at the end. For her, the question was so pressing that she announced her intention to retire from sailing in 2009, launching the Ellen MacArthur Foundation the following year. The organization has the stated mission to "accelerate the transition to a circular economy."

Along with the foundation's chief executive, Andrew Morlet, who left a twenty-two-year management consulting career to work with Ellen on this mission, she and her team have raised a towering flag on the importance of the circular economy movement. They have worked with globally influential businesses, governments, and nonprofits to not only point out the structural deficiencies of our system but also catalyze the scaling of solutions for those deficiencies. Building upon the work of those before them, the Ellen MacArthur Foundation isn't just accelerating the transition to the circular economy; they are leading it.

They have developed a great framework for understanding how materials will flow in a circular economy, which they visually depict as two complementary series of loops, one biological and one technical. Ellen and Andrew casually refer to it as their "butterfly diagram," since the two loops arranged horizontally resemble the wings of a butterfly. I am rather fond of the overt reference to nature.

Take a close look at their graphic (page 199). Materials flow from top to bottom, with the goal of preserving natural capital in our virgin stocks of material at the top and minimizing material that leaves the system as waste at the bottom. So long as molecules loop either in the biological cycle on the left or the technical cycle on the right, circularity is working and the take-make-waste linear sequence is broken. The graphic elegantly depicts the wide range of how materials can loop, depending on the material type and the stage of the economic process. Of critical importance, the circular economy is *not* about recycling. Recycling is simply one practice within the circular economy, represented by a single loop type in the technical cycle. We need the other looping practices as well.

Moreover, tighter loops in the graphic represent more efficient forms of circular economy practices. Consider this example. If you have an extra loaf of bread that you are not going to eat before it spoils, you could just go ahead and compost it right away. The carbohydrates would break down and create rich soil, which could follow the loop through the biosphere and back to farming and the harvest of a new wheat crop. The better solution would be to sell or give the loaf of bread to someone who would eat it before it spoils, which displaces the need for one extra loaf in the system. This practice is represented by the cascades loop in the biological cycle. We can intuitively understand why this action is better, since a loaf of bread is more valuable as food than compost, even though composting is a circular economy practice. The circular economy is fundamentally based on optimizing the value of materials in the system.

I have one last observation about the foundation's butterfly graphic, in particular when compared with Ray's Prototypical Company of the Twenty-First Century graphic. Do you notice how the Ellen MacArthur Foundation provides intricate detail about the systemic cycling of materials, but represents individual manufacturers and service providers with simple boxes? In Ray's graphic, the component parts of a company and its environmental and social linkages are clearly spelled out, while the biological and technical loops are represented by simple lines with little detail. The Ellen MacArthur Foundation provides the details that Ray glosses over, and vice versa. They represent the ideal depictions of two sides of the same coin. Ray's graphic depicts microeconomic circularity, with the focus on a single company, while the butterfly graphic depicts macroeconomic circularity, with a focus on material flows in the system. Just as understanding classical economics requires study of micro- and macroeconomic theories, understanding the circular economy requires a study of how companies within the system should operate and of the system on the whole. In other words, we have a lot to learn from both Interface's model and the Ellen MacArthur Foundation.

Systems Change

Sitting in the Roundhouse in London, I felt like a toddler whose complete attention is fully captured by every new toy put in front of him. A synthetic biotech advocate explained the emerging practice of cellular agriculture, which seeks to create lab-grown agricultural products with dramatically fewer resource inputs and associated waste. A digital expert explained how blockchain could be used to track information about the material makeup of commercial products. Blockchain is a digital technology that assures the validity of cryptocurrencies, originally developed for bitcoin, a digital ledger that allows users of a currency to trust that a unit of that currency is legitimate (kind of like watermarks in a dollar bill). If we could use blockchain to authentically track materials, we would be able to know exactly where valuable materials are in the flow of commerce.

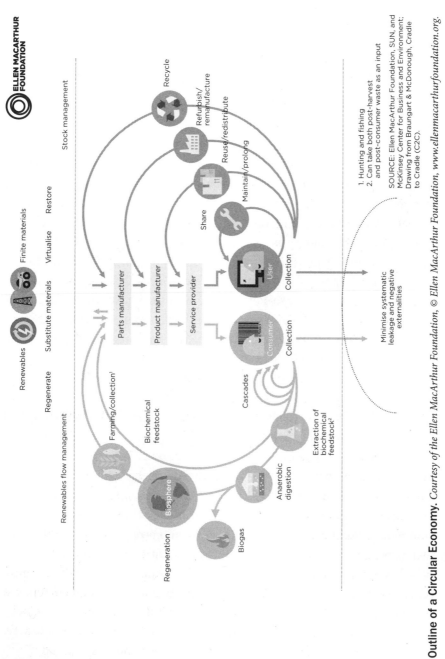

Outline of a Circular Economy. *Courtesy of the Ellen MacArthur Foundation, © Ellen MacArthur Foundation, www.ellenmacarthurfoundation.org.*

Renewables flow management

Renewables   Finite materials

Regenerate   Substitute materials   Virtualise   Restore

Stock management

Recycle

Refurbish/
remanufacture

Reuse/redistribute

Maintain/prolong

Share

Parts manufacturer

Product manufacturer

Service provider

User

Collection

Consumer

Collection

Farming/collection¹

Biochemical
feedstock

Cascades

Extraction of
biochemical
feedstock²

Biosphere

Anaerobic
digestion

Regeneration

Biogas

Minimise systematic
leakage and negative
externalities

1. Hunting and fishing
2. Can take both post-harvest
   and post-consumer waste as an input

SOURCE: Ellen MacArthur Foundation, SUN, and
McKinsey Center for Business and Environment;
Drawing from Braungart & McDonough, Cradle
to Cradle (C2C).

Later we would hear from an entrepreneur who started an online consignment business for luxury goods and another who was trying to automate insect farming with food waste as a feedstock. An expert on social change discussed how to create paradigm shifts, and another expert on political science argued that democratic deliberation is an important tool for advancing environmentalism.

Clearly, there are many different manifestations of the transition to a circular economy in tremendously diverse sectors. The Ellen MacArthur Foundation itself has initiatives focused on plastics, fashion, and food, and they also work at a broader systems level, understanding that systems change is necessary to transform isolated sectors.

"We talk about the broader economy as a system," Andrew Morlet said, "recognizing interrelatedness of all the different ways in which the economy can be characterized, and we come in through a materials lens. That's the core of the foundation's thinking—materials are an entry point into the broader economy as a complex system."

Understanding the economic system and its flaws is crucial, too. It allows us to envision what would be different, what would be better. If we want to go from Point A to Point B, we need to know where Point A is and where to go to get to Point B. As Ellen MacArthur told me, "It's about saying, *Where are we trying to get to? How are we going to get there?* Well, that's going to involve educators. It's going to involve research universities. It's going to involve the biggest corporations in the world . . . There is a huge amount of analysis and insight and conversation that goes into getting to that point of success, but it all boils back to that question, *What does success look like, and how are we going to get there?*"

Success does not look like Interface approaching circularity with a handful of other companies here or there. For certain sectors to change at all, they must change en masse. Remember my discussion in chapter 8 of Interface's reliance on the fossil-fuel-based transportation sector? Interface cannot make the transportation of its people and products sustainable on its own. Explained Jim Hartzfeld, "For Interface, there were not enough entities in the system who were engaged in the idea to catalyze the shift of the whole system. We were just one tiny part of that system as a billion-dollar company. What Ellen MacArthur has been able to do is capture the attention of massive portions of the industrial sector, agricultural sector, fashion sector, et cetera, so that all the players in the system are at least exploring this right now."

I am reminded of a surfer on a wave. That surfer can have the best board and world-class technique. Without a big wave, though, there is only so much that surfer can do.

To Joel Makower, "Circular economy is the most interesting thing going on in sustainable business. It is the first framework that I have seen that has caught on beyond a handful of companies. It requires a company to engage its entire value chain. It's not something you can just do by yourself, and you often have to engage your competitors."

I can offer a simple example of the value of this cooperation, one that I bet has frustrated every single person who is reading this. Have you ever asked a friend if you can borrow a phone charger, only to find out that they use a brand of phone with a different plug type from yours? It would be so much easier if Apple, Samsung, and the rest could just agree on a universal charger. If they did, we would not need extraneous charger converters or a portfolio of charger types for our various devices. We could use a lot less stuff to get what we want—fully charged devices.

In my mind collaboration is the key to the success we envision. In transitioning to the circular economy, we need collaboration between competitors, within supply chains, with consumers, and among businesses, governments, and NGOs. That will be the defining word of the systems change that creates the circular economy. That said, the transition to the circular economy will also involve specific types of transformations, such as better product design and material science innovations. These are the new practices that will bend the straight line of our linear economic system, curving it closer and closer to a circle. They are also the practices that will make the circular economy one of the greatest business opportunities our world has ever seen.

Bending the Line: Better Design

If you want to keep material from becoming waste, a good place to start is to keep material from becoming a product in the first place. This circular economy practice is called dematerialization, and it comes straight out of the Interface QUEST playbook. By creating the same-quality product with fewer raw material inputs, a business saves money and reduces demand for materials that might end up becoming waste. Dematerialization comes down to being as efficient as possible with what goes into a product.

It is a design challenge—one that Interface's principal designer, David Oakey, took up. When Ray had his epiphany, Interface carpet tile had at least twenty-four ounces of face fiber per square yard, and often quite a bit more. David and his team set their sights on driving that number down without sacrificing product performance. In the following years, they have been able to strip out more than four ounces per square yard, saving not only material but also the energy that would have been used in manufacturing the superfluous nylon fibers. On top of that, the weight savings generate subsequent fuel savings when Interface products are shipped. David's hard work generated multiple wins for the company.

Another recent example comes from Starbucks. In 2018, in response to growing public awareness about the massive quantities of single-use plastic straws that are used and disposed of every day, Starbucks announced their intent to stop using plastic straws by 2020. In their place, Starbucks has redesigned the lid they use for cold drinks, making it look like an adult "sippy cup." While the new lids use slightly more plastic, the extra plastic is still less compared with the plastic straws, and they are easier to recycle. Less plastic is used for effectively the same user experience.

Admittedly, it would be even better if customers brought in their own reusable cup and skipped a single-use cup in the first place. That point brings up an important observation about a lot of circular economy practices. Just because they are better, that does not make them the best, so we must keep working on improving. Still, marginal improvement helps to bend the line.

Dematerialization is also about preparing for a future with rising input costs. In *The Big Pivot*, Andrew Winston observed that while commodity prices generally declined in the twentieth century, they have tended to increase in real terms this century. In his words, "Commodity prices, while volatile, are now fundamentally heading higher . . . The imperative to reduce material use is rising. The only practical path for countries and companies that want to keep functioning profitably, or at all, is radical efficiency."[2]

Circular economy product design is not just about designing for less, but also about designing for long. Resource needs are reduced when a product can last longer and can be used more times. I will grant you that many consumable goods cannot be designed this way. The last time I checked, I only get to drink a glass of wine once. Durable goods can be designed this way, though, and so we bend the line toward circularity when products are designed for durability and repairability.

A Conversation with
Ellen MacArthur and Andrew Morlet

Ellen MacArthur is the founder and namesake of the Ellen MacArthur Foundation, and a retired English sailor. Andrew Morlet joined the Ellen MacArthur Foundation in 2013 to define and launch its business programs and became chief executive in 2014.

JOHN: How did you come to have a passion for the work you do?

ELLEN: For me, I guess I was the most unlikely person to end up in this space. I never thought I'd do anything other than sailing. So it was a real surprise to me when I started asking questions. When you're sailing around the world, you have a clear understanding of what finite means, because the resources you have on the boat are all you have.

But when it came to the finite resources of our economy, it just wasn't something being discussed in every country . . . even though the entire global economy's future depends on it. The more I learned, the more I was shocked how little this topic was being explored, and how limited the solutions were.

The initial question was, *How do we use resources?* If you ask that question, it quickly becomes clear that we use them in a very linear way, and the approach for future prosperity was to try and make those resources last a little bit longer. So my question then became, "If we are currently buying ourselves time, then what really works in the long term? What does success look like for our global economy?" Because buying ourselves time isn't success.

Once you get into this, you start seeing opportunities everywhere. It's your entire life, it's your world, it takes over your thinking space. I'm not someone who can do half a job.

ANDREW: The first time I became aware of Ellen was in 2001 when she was sailing the Vendée Globe race. I was actually working at that time with McKinsey in Korea. I had been a competitive sailor for many years, and Ellen popped up on the radar as a twenty-four-year-old out of nowhere. The Vendée Globe was the pinnacle of extreme sailing, single-handed around the world in a sixty-foot boat. It was brutal—and she came second by a whisker. Roll forward, I was a management consultant for twenty-two years and I worked for the world's leading management consultant companies all around the world, essentially optimizing linear, global company systems, and was very much in the world of efficiency and globalization, global supply chains, that whole space.

All the while, while I was doing that, it always nagged at me that there was an economic model that was about economic growth, quarter-by-quarter growth, that was massively consumptive and there was a rising world of evidence of environmental damage. I saw companies starting to look at efficiency issues and agendas that delivered sustainability as a by-product. So whether it was about green data centers or driving supply chain efficiency—which, after all, reduces impact and saves money, a dual benefit—it played well to a corporate agenda, whether you were focused on the environmental sustainability piece or efficiency.

Then I stumbled across Ellen again in 2012 when I came across a job advertisement. I called up and said, "This is interesting." The more I got into it, the more the idea of a circular economy and what Ellen explained really captured my imagination. Here was an idea that actually made a lot of intuitive sense and allowed you to talk about a compelling vision to drive toward, that was about innovation.

What is the role of nonprofits—yours and others—and governments in this space? Oftentimes systems struggle to find synergy between NGOs, governments, and the businesses that really need to be the ones making the change.

ELLEN: The shift to a regenerative, circular economy will touch all aspects of our economic system, and there's going to be common ground. There will undoubtedly be benefits that appeal to business, governments, and citizens. Another area for crossover is around collaboration. Working on the circular economy, it becomes very clear that you need collaboration like never before. You need agreement on policy, but agreement that keeps the vision and principles of a circular economy in mind.

It's about saying, "Where are we trying to get to? How are we going to get there?" That's also going to involve educators. It's going to involve universities and research. It's going to involve the biggest corporations in the world who say, "This is where we want to get to. How are we going to do that?"

ANDREW: Given that we're trying to shift complex systems, what we are focusing on is mobilizing industrial actors around this. Harmonizing global policies is a tricky thing to do; 190-plus countries would need to agree. We know also that the global economy could potentially double twice by 2050. We need to find a way of rapidly accelerating the transition to more regenerative, circular economy. We believe that companies, businesses, are the innovation engines of the economy and they play a crucial role in the redesign of the system, and we need to have policies emerge that are enabling that to happen using a variety of levers.

There is a third pillar of thinking here. Learning the difference between a circular and linear economy all the way down to teaching and providing higher education and university-level expertise for the designers and the technicians and the businesspeople that will innovate these models over the next twenty, thirty years. For us it's industry-led, policy-enabled, and includes the need to educate and inspire a generation to bring this about.

Who are your sources of inspiration . . . Who do you consider your teachers on the path that you've walked?

ELLEN: My grandmother got a degree when she was in her eighties. It's what she wanted to do all her life. That was a huge inspiration for me because she taught me: If you have a goal, then you need to somehow make it happen. When it comes to the circular economy, I'd say inspirations come from many areas. I live on a farm and I learn from the farm. It's a system that has its own flows of energy and materials, and cycles of regeneration. Even just walking down a road and seeing a chip packet stuck up against a fence and knowing it's going to be there now and in five years and fifteen years if nobody moves it away. All of these things make me think. They make me question, and hunt for answers.

ANDREW: We've sought inspiration in very broad ways, and it's not only about the theory. It's about how do we apply it, how do we actually learn and apply it in a way that can scale exponentially? We look very broadly in terms of inspiration, not only for the basic idea and how we continue to develop that, but also how we as an NGO are really different from the outset, because what we have seen is a lot of really well-meaning theoretical and small examples, but the magnitude of the challenge that we are facing is an exponential challenge and it is global.

We have an organization that's made up of extraordinary, diverse backgrounds, ranging from theoretical physicists and mathematician through to learning experts to communications experts across twenty-seven nationalities. We've built a group that is very different and we basically say to ourselves, "If the approach that we're adopting feels familiar and it feels like it's the way that we've always done it, then we've probably got the wrong answer."

Do you have thoughts on how capitalism as the dominant economic system is contributing to the problem of keeping us entrenched while at the same time its strengths might prove to be an accelerant toward a circular economy?

ELLEN: I go back to one of the first questions we ever asked McKinsey, when we were writing our first report in 2011: Does the circular economy have the potential to decouple growth from resource constraints? That's even acknowledging that we live in a predominantly capitalistic society. Actually, the answer came back, "Yes." The second question: Is it good for business? Is it profitable in business? "Yes." Now, is it good for the wider economy from an employment, general growth, prosperity perspective? "Yes, it is, because you do have the ability through the circular economy, through circularity, to decouple growth from those resource constraints."

Of course, there are many issues way outside circular economy. There's fair trade, there's fairness in the workplace, there's massive inequality in the world. We don't pretend that we—or the circular economy—has an answer to everything. But for many of today's most pressing challenges, it provides a vision and approach for a thriving, prosperous economy.

Patagonia is an exemplary company when it comes to designing products for the long haul. In his memoir *Let My People Go Surfing*, Patagonia founder Yvon Chouinard wrote, "As individual consumers, the single best thing we can do for the planet is to keep our stuff in use longer."[3] To encourage the repair rather than replacement of their products, Patagonia has a garment repair facility in Nevada, an archive of most fabrics and trims they have used, and online repair guides for their goods. The company wants customers to be "owners" rather than "consumers," and Yvon Chouinard emphasized the difference. "Owners are empowered to take responsibility for their purchases—from proper cleaning to repairing, reusing, and sharing. Consumers take, make, dispose, and repeat—a pattern that is driving us toward ecological bankruptcy."[4]

Many CEOs might shudder at the thought of discouraging customers from making an additional purchase from their company. I think this reaction is shortsighted, though. Durable goods can command a price premium, and offering repairability for a fee can diversify revenue streams for the business. Encouraging customers to trade in their products, as Patagonia does, can also allow businesses to generate multiple sales from one product. Further, as more and more customers become sensitive to environmental issues, companies that have an environmental ethic have a market-share advantage. Finally, when customers need to purchase a new product, they're more likely to return to the company that has provided them with high-quality goods in the past.

When simple repairs are not possible, the next best option is to remanufacture or refurbish durable goods. This is also a design challenge, in that both products and business models can be designed to make refurbishing easier. Doing so reduces energy and material demand, just as Walter Stahel observed in 1976. This practice can be particularly valuable for businesses in the electronics industry. Hardware and software advancements are incredibly rapid today, so refurbishing devices with the latest capabilities offers an opportunity to sell an even better product than the first one that was sold, while still utilizing many of the original components.

When even the most durable product reaches the end of its useful life, we need it to be designed for disassembly. Doing so would be a significant enabler of the recycling industry, which needs material types to be properly sorted in order for them to be recycled. Plenty of products end up in landfills simply because the plastics and metals cannot be easily pulled apart. There is opportunity for value creation here as well, as a lot of high-demand materials could have resale value if producers and consumers could easily disassemble products.

From dematerialization to durability to disassembly, products can be designed for the circular economy. All of this comes with a hidden benefit as well: It helps enable the transition to renewable energy. Let me connect the dots, with Andrew Morlet's help.

"Think about an engine in a car," Andrew said. "You have a massive amount of embedded energy and resource intensity in that engine, everything that was used to make it. If you keep it in use longer and you can remanufacture it, you retain that energy. Embedded energy is all around us, and the more you can keep an engine, a phone, a table, or a building in use, the more of that embedded energy that stays in use. You lower the demand for energy in the system."

This logic is similar to that for energy-efficiency improvements. From a global warming standpoint, reducing the need for fossil fuel energy through efficiency improvement is effectively equivalent to replacing the fossil fuels with wind and solar power. From a financial standpoint, many energy-efficiency improvements have a faster return on investment than wind and solar. That makes energy efficiency an attractive renewable energy investment opportunity, and the same can be said for the circular economy. Capturing the value of embedded energy could be a new frontier for our energy-intensive economic system.

Bending the Line: Material Innovations

John Picard has been busy since the first release of this book. In addition to advising Ray and Interface on their journey up Mount Sustainability, he has continued to work with other companies to advance their environmental efforts. I do not know anyone who is more willing to look around the corner for the technology of the future. Right now, Picard is captivated by the potential of a relatively new and cutting-edge material called graphene.

Graphene is a one-atom-thick lattice of pure carbon with remarkable properties. It is lightweight, incredibly strong, transparent, flexible, and very conductive. Its potential applications are near endless. It could be an additive to concrete, reducing the amount of aggregate needed for the same structural support. It could help make lightweight airplanes, allowing them to be more fuel-efficient. It could also theoretically be used to improve battery technologies.

"Graphene could help create supercapacitors, a sort of ultra-battery," Picard said. "It's the early days, but my prediction is that the ability to store energy is going to go 10^2. We could see a 10,000 percent increase in power density and storage."

The challenge is that graphene does not exist in nature. Creating it is rather expensive, but Picard is hopeful that breakthroughs will soon allow graphene to be commercially produced, unlocking its practical application. Massive investments are currently being made in creating materials of the future, whether graphene or otherwise. If you want to talk about business opportunity, the winners of the race for tomorrow's materials technologies will thrive.

The race is on in the plastics space as well. Reducing plastic pollution will require a large number of systemic changes—from consumer behavior to recycling capabilities to product designs—in addition to materials. "Some of the largest producers of plastic are digging deep into the circular economy," noted Joel Makower. "Their products aren't yet cost-competitive with some of the commodity molecules that they sell, but they understand where they need to go, and there's an incredible amount of innovation going on behind the scenes."

Could our plastics of the future be bio-based with ease of recyclability and a rapid breakdown cycle, all with no harm to the environment? Possibly. A lot will need to change, but there is significant opportunity to do well and do good while shifting this environmentally harmful industry.

Innovation around waste streams is also accelerating, and in the coming decades our durable goods and landfills could become raw material mines. "The concentrations of usable material per ton is much higher in manufactured goods than it is in a mine, and a lot of the material has already been refined once," Joel said. "Mining landfills for their increasingly valuable constituent parts is not only possible, but in some cases desirable to get required materials. That said, the innovations in sorting, assaying, smelting, and all the things that have to happen to separate tin from aluminum and gold from silver are yet to come."

Interface has already ventured into waste streams in search of raw materials. Since 2012 it has partnered with its yarn supplier Aquafil and the Zoological Society of London on a program called Net-Works. The Zoological Society of London works with local communities in the Philippines and Cameroon to gather old and discarded nylon fishing nets, which are then sent to Aquafil to be recycled into nylon fibers. Interface then purchases these recycled fibers for use in new carpet. More than two hundred metric tons of waste fishing nets have been collected in the program, with more potential for growth. As fishing nets become carpet tile, these poor communities in the Philippines and Cameroon benefit from a new revenue stream. It is a great example of how bending the line toward circularity can have cascading benefits.

Bending the Line: New Business Models

When most people hear the term *sharing economy*, they think of companies like Uber and Lyft. But as Rachel Botsman and Roo Rogers pointed out in their book *What's Mine Is Yours*, this evolving form of business practice known as collaborative consumption has many models: "Swap trading, time banks, local exchange trading systems (LETS), bartering, social lending, peer-to-peer currencies, tool exchanges, land shares, clothing swaps, toy sharing, shared workspaces, co-housing, co-working, CouchSurfing, car sharing, crowdfunding, bike sharing, ride sharing, food co-ops," and the list goes on.[5]

While some of these practices involve a business and others do not, a few commonalities tie them together. Generally, collaborative consumption involves exchanges of goods and services between regular, everyday people, often within the same community. To the extent that a business is involved, it often serves as the digital go-between, providing the software capabilities that enable the exchange to take place. Additionally, one person usually has an underutilized asset to be shared with or sent to another person, whether it be a car sitting in a garage, a handbag buried in a closet, a chain saw rusting in the shed, or a few hours on the weekend that would otherwise be wasted. Botsman and Rogers sum it up by identifying four principles of collaborative consumption: "critical mass; idling capacity; belief in the commons; and trust between strangers."[6]

From a circular economy standpoint, the benefit of collaborative consumption lies in the idling capacity principle. As Botsman and Rogers explained, "If you are like most people, you may use a power drill somewhere between six and thirteen minutes in its entire lifetime . . . There are approximately fifty million drills in homes across America gathering dust. Ownership of a product you use for just a few minutes makes no rational sense."[7]

Just imagine how much material and energy (not to mention money) could have been saved if it were more common to share power drills in a community rather than buy one for home use. The same can be said for bicycles, automobiles, lawn mowers, and a whole host of other durable goods. Collaborative consumption can keep tremendous amounts of material out of the flow of commerce, thus advancing the shift to the circular economy.

As collaborative consumption practices become more and more common, the businesses that facilitate them will succeed. Just as important, it will threaten traditional manufacturers and retailers. If people begin to borrow, share, and rent durable goods instead of buy them, the revenue from sales will dry up. For many organizations, the shift to the circular economy is more about planning for business risk than capitalizing on a new opportunity. Either way, businesses should be paying attention.

Interface certainly is. Remember in chapter 1 when Ray talked about the Evergreen Lease? Interface hoped to broadly lease carpet rather than sell it, recognizing that customers inherently care less about ownership and more about the services of carpet: aesthetics, comfort, and cleanliness, for instance. This orientation to the service of a good, rather than the ownership of the good itself, is core to the evolving business models that advance the circular economy.

Unfortunately, the transition is not always a smooth one. In Interface's case the Evergreen Lease has been shelved for the time being, waiting for its day to come. It did not become common at Interface for a variety of reasons, which are worth mentioning here. First and foremost, while customers do not fundamentally care about owning carpet, they are used to owning it. Convincing many customers to be open to a lease was difficult, simply because it was different. Further, the depreciation schedules for leased carpet made the accounting difficult. In the commercial sector many customers were purchasing carpet from capital accounts in their budget, and they would not want to shift carpet to their operational accounts. Even inside the company, it was challenging to properly compensate a commissioned salesperson for leased carpet compared with sold carpet.

All of these challenges can be overcome as broader economic conditions shift toward circularity. Moreover, even though the Evergreen Lease was ahead of its time, Jim Hartzfeld challenged me when I referred to it as a failure.

"We did not write a lot of Evergreen service agreements, but what that audacious, challenging idea created was a huge amount of coverage," he said. "Some of the world's thought leaders were telling our story to everybody else." Jim also pointed out that Interface earned some significant sales from clients as a result of pitching the Evergreen Lease. While they were not willing to lease Interface carpet, they appreciated the depth of the company's environmental commitment as shown by it. Simply offering the lease was another way to earn the goodwill of the marketplace.

Bending the Line: Policy

Though I conclude with public policy and the role of government in the shift toward circularity, I do not mean to diminish its importance. The rules of our economic game will significantly contribute to, or impede, the shift toward circularity, whether at a federal, regional, state, or local level. While all levels are important, how you influence policy at each level is different. Even more important, public policy always involves a balancing of many varied and often conflicting interests. What is wise policy in one place might be foolish in another, whether for cultural, economic, demographic, or social reasons.

Because of this complexity, I will not be prescriptive in asserting what policies should be adopted. Instead, I want to give an example of a policy shift that is having an impact on the circular economy, demonstrating that businesses need to be aware of how circular economy policies can be an opportunity and risk for their organizations.

In 2010 California became the first state in the United States to pass a law mandating a carpet recycling program. As a part of that program, a 5-cent tax was imposed on every yard of carpet sold in the state, which has since risen to a 25-cent tax. The revenue supports the carpet recycling industry in the state in a number of ways, primarily as a subsidy to recyclers who take back old carpet. This policy helps make carpet recycling profitable for the recyclers, therefore increasing the supply of recycled raw materials for carpet manufacturers.

In 2017, unsatisfied with the progress of the program, the California legislature strengthened it, causing a rift among manufacturers. As a result, the tax will rise higher and the funds will no longer be available to certain applicants, such as waste-to-energy incinerators. Interface and Tandus Centiva, one of Interface's competitors, came out in support of the policy change. Others in the industry opposed it. Dan Hendrix shared with me why Interface was in favor of the stronger law: "For us to get to our sustainability goals, we have to have recycled content available to us. A lot of these cottage-industry, mom-and-pop recyclers went out of business because there was not enough money. When you have a tax, it gives them money for recycling, so more people will get in and we will get more recycled content back for our products."

For Interface, the new law is aligned with both its mission and its business operations. Other companies see it purely as a threat to their sales. What happens if laws like these, generally called extended producer responsibility laws, become common in more and more states and more and more industries?

The clear answer is that businesses that are not proactive in becoming more circular will suffer financially. Those that have begun the shift will likely reap the benefit. Economics may not be a zero-sum game, but when one business loses market share, another one gains it. If I had to bet, such laws are very likely to increase in frequency as materials become scarcer and consumers become more aware of how environmentally harmful waste can be.

Something to Run Toward

Will some businesses work toward the circular economy simply because of a profit motive, without a concern for the environment? Perhaps, but does that matter? Said Ellen MacArthur, "Even if it's just about profits, then there is a reason to be in the room. This isn't about saying, 'Now, we all have to pay attention to circular economy because if we don't this bad thing is going to happen.' This is an opportunity. This is about innovation, creativity, reinventing business models. I think that there's an energy in the room. There's something to go for. There's something to run toward. There's an opportunity to be grasped with both hands."

CHAPTER TWELVE

A New Perspective on Our Climate

One afternoon in the summer of 2004, I was visiting with Ray and Pat in their home. I do not remember the occasion, and I suppose we did not need an occasion, but I do remember the year clearly. Ray had given me a high school graduation gift—a trip to San Antonio, Texas, for the NCAA Final Four. We had traveled in early April, just the two of us, to cheer on our beloved Georgia Tech Yellow Jackets in their last two college basketball games of the year.

I treasured the time we spent together that trip. We learned so much more about each other, and for the first time I was able to glimpse the depth of his passion for sustainability and Interface's mountain climb. I believe that trip was the seed planting my own environmental ethos, which is perhaps why one question in particular had been nagging at me for weeks. Sitting in Ray and Pat's living room, I got to ask it: "If you could wave a magic wand and fix just one environmental challenge, what would it be?"

I will always remember how quickly he answered with these two words: "Global warming."

Reflecting back on that long-ago conversation, I realize now just how correct Ray was. As he explained to me, global warming will impact the habitability of our planet. Certain regions will become too hot, too dry, or too submerged for humanity to continue living there. Such impacts make our warming planet a tremendous challenge.

Ray was also right for a more significant reason—global warming causes, amplifies, or is at least connected to nearly every other environmental challenge that we have. It is the ultimate "umbrella" issue. Consider biodiversity loss, for instance. Warming oceans are crippling coral reefs, the backbone of many aquatic ecosystems. At the same time, forest ecosystems are being decimated in the search for new agricultural land, not only releasing tons and tons of carbon dioxide in the process but also disrupting water cycles and both aquatic and terrestrial wildlife habitat.

Take the need for clean water, as much of an environmental concern as it is a human concern. Global warming complicates this as well. As storms become more severe, we will see even more impact from flooding as it overwhelms city sewer systems, breaches coal ash ponds from power plants, sends factory farm runoff into waterways, and otherwise taints local water supplies. In wet regions where mosquito-borne diseases are common, increased rainfall due to global warming could create more breeding grounds for these insects.

With examples like these in mind, I want to ask you what might seem like a silly question: In a warming world, what will we feel? I assure you, I am not being flippant. I know the obvious answer, and certainly many of us on planet Earth will struggle with heat waves. On a deeper level, though, I hope you see that the answers to this question are incredibly broad and unsettling.

In a warming world, a single mother of two small children might feel hunger. Already struggling to make ends meet and with her air conditioner working overtime, perhaps her energy bills will force her to make the gut-wrenching choice between electricity and food.

In a warming world, an elderly man from an island nation might feel the anxiety of settling into a new life in a strange country. He and other climate refugees will be driven from their homes and native lands as sea levels rise. Wherever these unfortunate souls settle, I pray that they are welcomed, accepted, and supported by their new countrymates.

In a warming world, a farmer who has labored on her family's lands for decades might feel fear. As weather patterns shift and her soils turn to dust, her prospects for economic opportunity simply blow away. How will she support her children when farming is all she has known?

Global warming is not only the most pressing environmental issue of our time—it is the most pressing *human* issue of our time. People all across the planet will feel the impacts of global warming, and many already are. How right Ray was back in 2004, and how desperately I wish that he had had that magic wand.

The world of business and industry is not exempt from global warming's transformative and destructive impact. This list, for instance, just skims the surface of how a warming world will challenge our economic system and the actors within it:

- How will the fashion industry respond when cotton crops routinely fail due to excessive heat? Costs will increase as the supply of cotton falls, and switching to alternative fiber types will also be more expensive as their demand increases.
- Similarly, are businesses in the food, beverage, and agriculture sectors prepared for shifting crop zones or for harvests being ruined by heat, drought, or extreme weather? Food is as much big business as it is human necessity in this modern world of ours.
- What will the insurance industry do as natural disasters increase in frequency and severity? Sure, they can raise rates to reflect higher risk factors, but there are limits to what people will pay. Eventually, customers will forgo insurance or move.
- Plenty of countries rely upon revenue from the tourism industry, which will have to adapt in a warming world. For instance, how will the coastal towns of Queensland, Australia, so reliant upon the majesty of the Great Barrier Reef, sustain themselves when the reef bleaches entirely?
- As global warming magnifies unrest in the destabilized countries of the world (likely due to more frequent droughts), what will the subsequent violence and mass migration do to the supply chains of all sorts of industries? The disruptions will manifest in many harmful ways.

It's not difficult to imagine trillions of dollars at stake as global warming is felt in its myriad forms throughout the world's industries. Still, despite the ominous clouds on the horizon, many businesses do not yet acknowledge how global warming will affect them.

Andrew Winston, as a consultant to large corporations on environmental issues, has witnessed this lack of foresight firsthand. "I have worked with hospitality companies and real estate companies," he said, "but I haven't seen any of these big companies with lots of assets do a systematic analysis of their coastal assets. They don't stop and ask, 'Should we really build a new resort in Miami Beach?' I think the risk side gets a lot of attention, but I'm not sure that many companies and people have really thought it all through."

It would be easy to assume that denial in the business community is the reason for its collective blind eye toward our climate. But I do not believe that is the case. While certain businesspeople might personally deny the scientific consensus on global warming being caused primarily by humanity, few if any global businesses take that position publicly. Even fossil fuel companies understand that the science is settled. So why is more climate action not being taken?

Joel Makower has a compelling answer to this question. "The problem across a lot of companies is that climate change is number four on the list of top three environmental problems they are facing," he said. "They are dealing with resource constraints or the right to operate in communities or a consumer pushback about wastes or toxins or the lack of repairability of their product. At what point does climate become the number three, two, or one issue? That is not yet the case in most companies."

The blind eye, it seems, arises from the short-term focus of the global economy. Our warming planet, which has been sending us warning signals for many decades, still seems like a threat for the future to address. When a company is balancing the environmental concerns Joel listed with meeting payroll, conducting the next board meeting, evaluating a proposed acquisition, and ensuring compliance with a wide range of regulations, how can it be expected to worry about the climate? The solar panel manufacturers are taking care of that, right?

They are not, despite best efforts, because the challenge is too great for one sector to solve. With only a few rare exceptions, every business on Earth has a carbon footprint, meaning it is a net emitter of greenhouse gases as a result of its business operations. Those carbon footprints are adding thickness to the atmosphere, our planetary blanket. Just as with the blanket on your bed, the warmth will continue to increase over time. Those carbon footprints make business and industry complicit in the threat to our climate's stability.

Blame, however, does a poor job of inspiring change. Moreover, castigating business and industry for their role in negatively changing our climate sets the wrong tone. It would suggest that our goal should be to simply "stop a bad thing"—basically the carbon loading of the atmosphere. The ultimate goal of such discourse is almost always to avoid climate catastrophe. I, too, want to avoid climate catastrophe, but I also want more than that. I want society to take the next step, evolving to a more just, equitable, prosperous, and resilient version of itself. When we advocate for climate action, shouldn't we aspire to more than simply avoiding catastrophe?

If you agree that the answer is yes, and I earnestly hope that you do, then what language and messaging should we use to aspire to this higher goal? I would suggest messages of belief, positivity, collaboration, aspiration, and opportunity. That is the language of this book, beginning with Ray's first keystrokes two decades ago. Its usefulness has been validated in the hundreds of speeches that Ray gave advocating for sustainability. People responded to his message of doing well by doing good. They will also respond to the vision that by reversing global warming, we will create the best version of humanity that has ever existed. I can think of no other sector better suited to champion such a vision than business and industry.

Do you find yourself responding skeptically to what I have just written? I understand if so. We all know we are confronting challenges that some have even called insurmountable. I am not asking you to disregard dire projections. We need to be conscious of the problems we will encounter so that we can appropriately prepare for the dramatic change that is coming. But we cannot let them occupy our full range of vision.

You also might be skeptical about the link between business-led climate action and the next leap in human prosperity. If so, let me paint the picture of what I believe our world will look like when we achieve climate stability. From an energy standpoint, every electron will come from a renewable source, with costs driven down to their bare minimum. This future energy system will make electricity accessible and affordable for everyone. From a food standpoint, systems will be localized with minimal waste and better human health outcomes as a result of plant-rich diets. They will also offer economic opportunity to more people, decentralizing the process of satisfying the basic human need for sustenance. From a land-use standpoint, wild spaces will be preserved and restored, and managed land will be optimized for soil health, generating better and more sustainable yields no matter the crop. As a result, natural systems will thrive, providing all of us with enhanced ecosystem services like clean air and water. From a transportation and built-environment standpoint, better community outcomes will flow from efficiency optimization and system redesign. Human equity will be a much-needed by-product of climate-friendly evolutions in these sectors.

Taken together, these transformations offer a vision for a more prosperous and resilient world than we have today. Now, can you imagine any of them taking place without business and industry leading? I cannot.

In case doubt still tugs at your mind, I expect it is because you do not see the connection between reversing global warming and creating this new future. I assure you it is there, and much stronger than you might think. The reason is simple—nearly every technology and practice of this hypothetical future yields a measurable reduction of atmospheric greenhouse gas concentrations as it scales globally.

As proof, I offer to you the tremendous work of Project Drawdown, founded by none other than Ray's first sustainability teacher, Paul Hawken. First, let me share a bit of the organization's backstory. In 2001 Paul was reading the IPCC Third Assessment Report on the state of global warming. He remembers wondering why, despite so much certainty about the science, there was so little discourse about the solutions to the problem. "I understood energy efficiency and renewable energy," he said, "although I knew it was very expensive. Beyond that, I thought, *There is nothing out there that gave me a sense that we could actually accomplish it.* I don't even think anyone was talking about what *it* was."

Paul started asking around. Is anyone naming the goal? Is anyone creating a list of solutions and what they could do? For the most part, the answers were "no" and "no." Occasionally, the answers were "no" and "why don't you do it?" In 2013 he finally decided to.

He began by naming the goal *drawdown*, defined as the point in time when atmospheric concentrations of greenhouse gas emissions peak and begin to decline on a year-on-year basis. Slowing the rate of greenhouse gas accumulation is not enough: To reverse global warming, we must see more carbon come out of the atmosphere each year than goes into it. Paul then founded the organization that would discover what tools humanity has to achieve this goal.

Simply put, Project Drawdown did the math. They mapped, measured, and modeled the most substantive solutions to reverse global warming, releasing them as a book titled *Drawdown* in the spring of 2017. The findings were rather surprising. For instance, eight of the top twenty solutions were in the food sector, including two of the top five in reduced food waste and plant-rich diets.[1] Solar farms and rooftop solar are both in the top ten, but combined they still have less carbon reduction potential than onshore wind turbines, the number two solution. Educating girls is a solution in the top ten due to the reduction in birth rates, primarily in the developing world. That solution is perhaps the best example of how our improved society of the future can both be more equitable and more climate-stable.

In addition to the carbon metrics, the team also modeled the financial impact of scaling these solutions, at least for every solution that could be modeled financially. Quite a few solutions, like indigenous peoples' land management and creating walkable cities, cannot be modeled that way. For accelerating the deployment of technologies like heat pumps and high-speed rail, on the other hand, the team was able to model the net cost, defined as the additional cost to implement these climate solutions compared with repeating business as usual, and net operational savings of scaling the technologies. The team found that, with the exception of refrigerant management and wave and tidal energy, every solution has a positive return on investment over the thirty-year scaling time frame. Altogether, Project Drawdown's models project $73.9 trillion in savings for a net cost investment of $27.4 trillion.[2] That thirty-year return equates to an annual rate of 3.36 percent, which isn't too bad considering that mature national economies routinely grow by less than that amount. Who knew that reversing global warming could be profitable?

In each of the previous two chapters, I have argued that we need to understand both microeconomic and macroeconomic aspects of our global economic system in order to change it as intentionally and effectively as possible. I want to make that argument once again, this time in response to the massive economic impact that global warming will have. In the climate space, Project Drawdown helps us to understand how our economy can shift at a macro level. First, their research and modeling show that reversing global warming is actually possible. I can think of no greater macrosystemic and economic consideration than whether we can achieve our desired goal. Plenty of people fear that we are already too far beyond our climate limits, but I do not believe that their fear is yet justified. When you add together the wide range of climate solutions that humanity is already implementing, the goal becomes achievable. To reiterate, though, we will fail if we look to one sector alone to solve the problem. We need an interdisciplinary approach.

Second, Project Drawdown does its work with a global systems analysis lens, and they built their model to account for the systems dynamics of solution scaling. For instance, scaling electric vehicles will increase electricity demand and impact the growth trajectory of electricity from renewable sources. As another example, converting degraded agricultural land to regenerative agriculture means that the same acreage cannot be

dedicated to afforestation. If the model did not account for these interconnections among solutions, then carbon benefits might be double-counted
and therefore overstated. This systems approach makes sense considering
that the warming is just feedback from our climate system. We are already
dealing with a systems-scale problem, so a systems-scale set of solutions
is required. As Paul told me, "When you see this system and this climate
feedback, you understand that our influence on the planet is extraordinary." Our influence thus far has been to the negative, but we are equally as
capable of having a positive influence on our climate.

If you flip through the pages of *Drawdown*, you will see the overarching characteristics of a climate-stable world. It is a world that I want for my
children and theirs, and it will help to bring about a macroeconomic system
that has a much greater chance of sustaining itself into the indefinite future.
Whether we successfully hitch our economic wagon to human capital or
we rethink the need for growth itself (the two sustainability pathways I laid
out in chapter 10), the low-cost energy sources and broad efficiencies of a
climate-stable world will be incredibly beneficial.

While macroeconomic in scope, Project Drawdown is also an invitation to everyone to participate in the solution to global warming. When
I asked Hawken what it will take to achieve drawdown, he was quick to
answer: "What it takes is one person after another getting engaged whether
they are a farmer or a mother or a father or a mechanic or a librarian or a
nurse or a teacher or a leader or a professor, it doesn't matter." The solutions
identified by Project Drawdown are so broad that everyone can find ways
to exercise their Power of One. In the context of business and industry, the
Power of One can be quite significant and quite beneficial, which brings us
to the microeconomic side of the equation.

The work of reversing global warming will be the sum total of millions
and millions of choices, actions, and transformations that either pull
greenhouse gases from the atmosphere or stop their emission into it.
We will approach success person by person, company by company, and
community by community. While the scale is daunting, we have good
news—there is tremendous opportunity in each and every climate-friendly
decision, and doubly so in the context of business and industry. I love how
Joel Makower put it in our conversation (see the sidebar in chapter 8):
"Climate change is a massive business opportunity masquerading as an
environmental problem."

In exploring this opportunity, I want to start by disclaiming one understanding of it. Global warming could be thought of as a "bubble," similar to the mortgage-backed security bubble of mid-2000s that caused the Great Recession. In that instance, real estate was broadly overvalued due to artificial demand stemming from fast-and-loose lending practices, and certain individuals with foresight decided to "short" the market by taking economic positions that would be profitable to them when the market experienced a downward correction. It did, and they certainly profited.

The prospects of global warming suggest that other asset types are similarly overvalued. Two examples are agricultural real estate in regions projected to become drier in the decades to come and coastal real estate expected to be inundated with floodwaters as sea level rises. I would argue that the securities of corporations with large carbon footprints are another example, since they will see rapid increases in the cost of doing business when governments and markets finally put a price on carbon.

While a downward correction of the valuation of these asset types does not yet seem imminent, it is coming. Could smart people short them and reap financial rewards? Could other smart people buy up asset types likely to increase in value as our climate warms, such as agricultural real estate in cool climates that will become more temperate? The answers are "yes" and "yes." Those opportunities do not interest me, though. They have to do with economic gain as our climate spirals out of balance, which in my opinion severely erodes the inherent value of such economic gain. What interests me are the microeconomic opportunities that come from helping to stabilize our climate.

I believe the opportunity side of climate action manifests itself in three ways, echoing the sustainability benefits described in chapter 9. I have already hinted at one in this chapter (though at a global scale)—the cascading benefits that come from climate action. Similar cascading benefits exist at micro-scale as well.

Steps forward are not limited to the kinds of product redevelopments or supply chain transformations we have discussed in past chapters. Take, for instance, a large corporation with a cafeteria in its corporate headquarters. If that cafeteria is like the ones I have visited, it tends to have lots of high-calorie and animal-derived protein offerings. What if they overhauled the cafeteria, offering exclusively local foods with an emphasis on plant-based options, while composting scraps and donating excess food to a nearby homeless shelter? Sure, the carbon footprint of the corporation would drop as a result of limiting animal products, sourcing locally, and composting. At the same time, employee productivity is likely to increase due to moderate calorie consumption, limiting the "afternoon crash" that I am sure we have all felt after a big meal. Employee health will also increase with the better diet, and the community's goodwill is earned with the donations to the shelter. The compost might offer a slight revenue stream, and it will also support the localized agriculture from which the corporation is now sourcing. As to the downside, employees might grumble about the more limited food choices, but with effective communication of these multiple benefits, the corporation can use the change to educate its employees about the environmental and social impacts of their diets, helping to shift the employee culture toward a higher purpose. Another downside could be the cost of such a transformation, but when you properly account for all of the benefits, I would bet that the benefits win out.

Other examples abound. By prioritizing telecommuting instead of air travel, corporations can reduce carbon while saving money and helping their employees spend more time with their families. By installing smart glass (windows that darken when the sun is shining through them), corporations can reduce carbon through lower heat loads while creating more comfortable work conditions for employees. By locating office buildings near public transit, corporations can reduce the carbon from employee commuting while minimizing the chances those employees will be in a costly or harmful automobile accident. The list could go on and on.

Positive climate action almost always increases the resilience and independence of an organization. Which source of natural gas would you prefer for your business—the one from the local capped landfill that is piped directly into your manufacturing facility or the one provided by a utility dependent upon the more volatile global market? The latter likely carries more risk as it is subject to potential market disruptions and regulatory change (just look at the environmental harm caused by fracked natural gas and consider whether that sector has exposure to the risk of regulatory change). Further, there is a good chance that the local capped landfill can sell the gas at a lower price.

Some businesses might even be able to achieve energy independence now or in the near future. Through the combination of on-site renewable energy generation and energy efficiency, companies that can divorce themselves from an external energy provider will have a strong competitive advantage. When a price on carbon finally arrives, helping the markets to fully internalize costs as any proponent of free markets should want, corporations with energy independence and low carbon footprints will thrive.

Finally, just as sustainability is an innovation catalyst, so, too, is climate action. I know of businesses today that are working to turn carbon dioxide into valuable raw materials, such as concrete and metal alloys. I do not yet know how successful they will be, but those attempts at innovation are in direct response to the challenge of global warming, and they could be remarkably profitable if the companies perfect their innovations. The race for next-generation battery technology highlights another example of an emergent technology at the intersection of climate and business opportunity. What other innovations are yet to come in response to our destabilized climate? While the solutions curated by Project Drawdown show that current technologies can reverse global warming, they are far from an exhaustive list when you consider the vast potential of the new technologies, Ray's T_2 technologies, that are yet to be invented. How can your business innovate new technologies or practices that reduce our atmosphere's carbon load? Perhaps your innovations would be unique to your business, but maybe, just maybe, they could revolutionize an entire industry or sector.

Would you like an example of a company that is committing to the goal of drawdown, hoping to lead business and industry in advancing bold and forward-thinking climate action? One that is proving how an authentic response to global warming is a better way to a bigger profit, just as Ray showed was possible in responding to sustainability? I do not have to look far. Once again, that exemplary company is Interface. This time, Ray is the spiritual leader of the journey, rather than the physical one.

As you now know, Interface is nearing the summit of Mount Sustainability. Even back in 2015, when Jay Gould joined Interface, time was beginning to run short on the Mission Zero pledge. Knowing how important that commitment was to Interface and Ray's legacy, Jay decided that fall to pull together several members of Ray's EcoDreamTeam for the first time since Ray passed away. "When I called them together," he said, "I was looking for an acknowledgment that we were going to come really close to reaching the top of the mountain by 2020 and delivering on Ray's commitments. Paul Hawken was stern with me, though, and he said, 'Look

Jay, that is good what you're doing, but Interface is the kind of company that takes on the world's greatest problems, and you need to do more.'"

At that time, Paul was already well into the process of creating Project Drawdown, so it should not surprise you that global warming was on his mind. Paul, along with Janine Benyus, Bill Browning, and John Picard, gave Jay a crash course on our changing climate. He came out of that meeting with a much deeper understanding of global warming but still quite unsure how Interface could tackle it. Despite the uncertainty, Jay ultimately decided to move forward and craft a new mission around global warming, and he gives Ray a lot of credit for why.

"It all was based on the progress that the company had made over the previous twenty-four years," Jay told me, "and the courage that Ray had demonstrated to say, 'We are going to go on a journey and we don't know how to accomplish it, but we know the world needs us to do this.' So it was with that backdrop that we were willing to take on climate change."

In a sense it was Mission Zero all over again. In 1994 Ray's epiphany revealed sustainability to be imperative, not optional. He committed Interface to a new course, trusting that his people would find a way. Jay came to see global warming as another business imperative, so he was willing to commit Interface again without fully understanding how the company would walk this new path. He and Erin Meezan, Interface's chief sustainability officer, got to work.

In the coming months they learned that Interface would have to execute on four concurrent strategies to be a company that was reversing global warming. First, it must "Live Zero," incorporating the decades of work that it had done to eliminate its negative impacts. Rather than a ceremonial hat tip to Mission Zero, they understood that Mission Zero's efforts are what have positioned the company to be an effective actor on climate. Moreover, as Interface grows as a company, perhaps even acquiring new subsidiaries, it will need to address any new negative environmental impacts that might arise.

Second, Interface must "Love Carbon," finding ways to utilize carbon dioxide as a resource. If carbon dioxide can become an integral part of carpet tile, then Interface carpet might become a carbon sink. This effort is arguably the most crucial one if Interface is to become carbon sequestering in its operations.

Third, Interface must "Let Nature Cool." Mother Nature already has a wide range of carbon sinks, and she is very good at sequestering carbon.

A Conversation with Paul Hawken

Paul Hawken is an environmentalist, entrepreneur, author, and activist who has dedicated his life to environmental sustainability and changing the relationship between business and the environment. He is a member of Ray's original EcoDreamTeam.

JOHN: You are the reason Ray's life changed, and it was your words that changed his life. I know he reached out to you not long after reading *The Ecology of Commerce*. I would love to hear your memories of meeting Ray and coming to know him.

PAUL: The first time he reached out to me was in the form of a letter when I was living on a houseboat. I had never heard of the company, or of him. In the letter he said he was challenged by his team for an environmental vision and he realized he didn't have one. He was given *The Ecology of Commerce* by Joyce LaValle, a vice president at Interface, and he humbly noted that he had copied parts of it for a speech to his employees about an environmental vision. That was the beginning.

How and why do you think the vision took hold at Interface?

I think the impetus within Interface to address the environment really first took hold in Ray—spiritually, emotionally, intellectually, and ethically. It took hold on all those levels, but then there was the implementation of it in a twentieth-century industrial company that was small by some standards but big by other measures.

There was initial skepticism within the company and fears about sustainability costing too much money, and there were worries about Interface being marginalized by Wall Street. Some people thought Ray might be seen as someone who lost his corporate way and got entrapped and captured by the environmental movement.

Nevertheless, Ray's vision took hold inside, and as more people became literate in sustainability it began to flourish. It started to energize the company and define it, as it moved from an idea to innovation and practice. And it wasn't long before the company started to get credit for its pioneering leadership. And that was due to Ray's steadfastness.

Ray showed that the path to sustainability was not a path of compliance but a path to innovation and a competitive advantage. Avoiding environmental breakdowns was actually a path to breakthroughs. And I believe most everyone got it. If there were doubters, they stayed well hidden.

What sparked Project Drawdown?

For a long time I was one of those people who thought climate change was huge but one of many issues—human health, toxicity, biodiversity, deforestation, and more. In 2001, when I read the summary of the Third IPCC Assessment, I didn't understand why we were talking almost entirely about the problem of global warming and not about the solutions. For years, I was watching how global warming was being communicated and seeing how the main communication tool was fear. Given the science, the fear is more than justified. However, fear is not a motivator; what happened is that it created "freeze and flight." Most people went numb, turned away, did not want to do anything, or felt disempowered.

I wanted to know where we stood. What were the solutions to global warming, all of them? So I launched an organization to map, measure, and model the 100 most substantive existing solutions to global warming and called it Project Drawdown. This identification and modeling had never been done before. Moreover, the goal had never been named: drawdown, that point in time when greenhouse gases peak and go down on a year-to-year basis. So we did both, named the goal and then collaborated with scholars and scientists around the world to see whether we could achieve drawdown by 2050 if existing solutions that were in place continued to scale in a rigorous way.

What has the response to Project Drawdown been?

If we had fantasized about the best case response, we would have underestimated what actually has happened.

Drawdown was an instant *New York Times* best seller and is published in twelve languages around the world. The drawdown concept is slowly but surely permeating the world, becoming a reference point for all types of organizations, including schools, universities, cities, investors, NGOs, countries, and more. Every problem is a solution in disguise, and global warming may be the most gnarly problem ever identified and defined. What we pointed out is that as a problem it is redolent in transformative solutions. It is creating optimism and hope. People who have read it acknowledged that they hadn't realize how much grief and despair they were carrying.

Collaboration is at the core of what we do; two-hundred-plus-some people from around the world created the model, and many of the scientists are millennials, which is kind of cool. It wasn't old people talking to young people—it's young people putting together the model and saying, *This is what we know.* There's no telling the impact it will have as it continues to scale. It's a very practical and grounded book that is being picked up by many organizations. Universities are talking about regionalizing the model.

We wrote a book that didn't blame or shame and didn't create fear or threat or doom but honored the science and identified the possibilities. It didn't demonize or say someone is wrong or bad or not good, and we actually didn't even say that we were right. We think it is approximately right. All models of the future are wrong, but some are useful.

Just like any good model. No model is exactly right; that is not the point.

What we have tried to do is create the conditions for self-organized groups around the world who want to address global warming. Top-down organizations are too slow and get in their own way. And it seems to be working. There is Drawdown Europe in Berlin, Drawdown Switzerland, Drawdown Netherlands, Drawdown Nova Scotia, Drawdown New Zealand, and many more.

How human beings change is mysterious. However, we do know that we move toward possibility and are not motivated by the probability of doom. What we did is essentially hold a mirror up to the world and revealed what we know and what we are doing. Drawdown shows, I think, that humanity is on the case and deeply cares.

She is our greatest ally in creating climate stability. Interface should therefore take positive actions to support the biosphere and enhance the environment's capacity to regulate climate.

Fourth, Interface must "Lead the Industrial Re-Revolution." Harking to the sixth and seventh fronts of Mount Sustainability, Interface knows the importance of its influencer role. The company cannot reverse global warming on its own, so it must constantly seek to bring others along for the journey.

Together these strategies make up Climate Take Back, which Interface officially launched as its next mission in June 2016. Importantly, these strategies are *what* Interface believes it needs to do. *How* Interface will accomplish them was not known at the time, and it did not need to be. Facing a challenge as daunting as global warming requires bold commitments, and we simply do not have enough time to wait for the "how" to become clear before committing to the "what."

Fortunately, Interface has a long track record of empowering its employees to discover how transformative goals can be achieved. The company is using the same approach with Climate Take Back, and we are already seeing positive results, especially on the Love Carbon strategy. John Bradford shared with me the initial thinking that he and his innovation team had when first challenged by Climate Take Back.

"We want to take carbon dioxide and turn it into a mineral-like form," he said. "There are a lot of things in the world that do that very well, and one is called a tree. It breathes carbon dioxide and turns it into a sugar. Our goal is to create processes to take carbon dioxide out of the air and put it into our product in the form of a mineral or plastic that will stay there for over fifty years."

Less than a year after Climate Take Back was launched, the company had developed a prototype carpet tile with a negative carbon footprint.

Specifically, the backing of that carpet tile sequestered more carbon than was emitted in the full life cycle of the product. In 2018 Interface announced that all of its products going forward will be carbon-neutral, admittedly relying in part on carbon offsets. Jay Gould and John Bradford believe that by 2020, Interface's products will not need these carbon offsets to be carbon-neutral, and even carbon-sequestering. Every sale of Interface carpet will mean that a little more carbon dioxide leaves the atmosphere.

Interface is reusing another strategy from its Mission Zero playbook here—biomimicry. In an effort to execute the Let Nature Cool strategy, Interface is working with Biomimicry 3.8 on an initiative called Factory as a Forest. The idea is simple—Interface wants its factories to generate all of the ecosystem services that the land would have provided if the factory had never been built, including carbon sequestration. Biomimicry 3.8 is helping by studying the reference ecosystems for Interface's facilities and developing strategies for how they can be redesigned. In the coming years Interface's facilities personnel will take the lead in implementing the most effective strategies. I am excited by the scope of the project, in particular for the ecosystems approach of biomimicry here, and I think it shows that Climate Take Back will be just as much of an innovation catalyst as Mission Zero has been.

In these early days of Climate Take Back, it is still uncertain in which direction the journey will take Interface. We do not yet know all of the challenges, the benefits, or the lessons that will be learned along the way. That said, I am willing to make a few observations thus far and predictions about what is to come.

First, the people of Interface are invested in this new challenge, which is likely the most important litmus test for its worth. Simply put, Climate Take Back cannot succeed without employee engagement, so it was wise of Jay and Erin to begin with broad outreach to their people. When they asked Interface employees if they were ready for a new mission, the answer was "yes" and "make it bold." Climate Take Back reflects their willingness to build upon Mission Zero's foundation.

Second, moving forward with Climate Take Back was not an easy decision, especially at the level of the board of directors. I asked Jay what the board's response was, and he said, "I think there was a lot of skepticism, to be perfectly honest with you, and the provocative language that we used was unnerving for them. But most of our board members joined because we are a courageous company, and joined because of the things we have done to build the value of our company while also driving sustainability. And so they were accepting, although skeptical."

Reversing global warming is as audacious a goal as can be set. Skepticism is a perfectly fair response, though I applaud the board for also being courageous. Without a board that is forward looking and willing to take smart risks, we will not see the transformative change that is needed on issues like global warming. I hope to see more CEOs cultivate such boards in the years to come.

Third, while Climate Take Back is underpinned by a strong moral ethic, Interface once again believes that it is tapping into a competitive business advantage. I assure you, Jay Gould is no less a fierce business competitor than Ray was. The advantage Interface is creating may have a long-term frame of reference rather than short, but we need businesses to be thinking this way. Moreover, if Interface is making a smart bet here, the payoff could be tremendous.

For instance, Interface believes that the carbon footprint of products will eventually enter into the building specification process. Just imagine if prospective clients always saw the carbon footprint of Interface carpet compared with its competitors. If Interface's competitors do not get to work now on eliminating their carbon footprints, they will be caught flat-footed. Interface would not win every bid, but it will win a lot more with its vastly better carbon metrics.

Interface also believes that in addition to customers becoming climate-conscious, politicians eventually will as well. "There is a day of reckoning coming," Dan Hendrix told me. "When there is a carbon tax, if you haven't built your supply chain and operations to reduce your carbon footprint, it isn't going to be easy to get started and catch up with Interface's twenty-five-year lead."

I strongly agree. The negative externality of global warming is not currently factored into the price of the products that we buy. Eventually there will be enough political will for a price on carbon to be legislated, helping to correct this market failure. When that happens, the businesses that have optimized around carbon footprints will be able to win in the market on both product performance and price. Interface's competitors have been warned! It is time to start competing to be the best at reversing global warming. Interface is out to a big lead.

Fourth and finally, Climate Take Back is an invitation. It is an invitation to Interface's competitors to compete on climate: If anyone can outdo Interface, we all win. It is an invitation to other businesses to collaborate on climate: Interface will happily learn best practices from others who are learning to love carbon and help nature cool. It is an invitation to all of us to care about climate: Our species has never faced a more critical challenge. Most important, it is an invitation to create a new global economy, one that is a friend to our climate rather than a foe.

We can do this. With drawdown as the goal, global warming will begin to be seen through the lens of possibility rather than problem. Stimulated by the feedback from our changing climate, humanity's brilliance will create the re-revolution needed to ensure a stable climate and a thriving species. At mid-century, I think we will be astounded at how well our economy serves us, rather than the other way around. We will look backward with awe and gratitude, and in the words of Paul Hawken, we will know that global warming did not happen to us. It happened for us.

CHAPTER THIRTEEN

More Ray Andersons

S unshine feels different at Serenbe, though it is difficult to explain why. If I had to guess, the towering pine trees have something to do with it. So, too, do the creaking sounds of rocking chairs that accompany friendly hellos as you walk through the community. From the farm to the livestock to the edible landscaping to the preserved greenspace, Serenbe is a celebration of Mother Nature's grace and beauty. When I turn my face toward the sun there, it seems as if Mother Nature is smiling with appreciation for this special place.

Serenbe is a thousand-acre community about thirty miles southwest of Atlanta, Georgia. Steve and Marie Nygren designed it around a deep sustainability ethic, though in the beginning they had no intention of creating this environmentally minded community. Having been successful as restaurateurs, they decided to sell their company in 1994 and move to the countryside. They enjoyed their quiet life with their three daughters, but soon urban sprawl began to creep to the edge of their property. Steve realized he would have to act to preserve the natural beauty all around him.

For decades Steve and Marie had been dear friends with Ray and Pat. One evening over dinner, Steve voiced his new concern over the approaching sprawl and asked Ray who might be able to help him stave it off. Ray didn't just suggest a couple of names. He made phone calls and brought twenty-three thought leaders to Steve's home in September 2000 to discuss how he and Marie could plan a community that developed with nature, not in spite of it. Serenbe was created with sustainability at its core, and Steve gives Ray great credit not just for his influence on the place, but also Ray's influence on him. Steve said to me, "It was that whole experience that Ray orchestrated that brought me from being a concerned citizen taking care of my family to understanding the environmental issues for the built environment."

Today Serenbe is thriving with hundreds of residents, a twenty-five-acre farm, an impressive array of charming stores and restaurants, and cultural assets like the Serenbe Playhouse. The community continues to grow, all while the majority of land is permanently dedicated to green space. It is an amazing success story, and Steve is kind to mention Ray anytime he tells it, but it was Steve and Marie who labored tirelessly to bring Serenbe from an idea into reality. Ray simply pointed them in the right direction, and then cheered them on for the remainder of his life. Serenbe may be an example of Ray's influence, but it is also an example of Steve and Marie's brilliance and hard work. In a sense Steve and Marie became Ray Andersons of their own.

I mean something very intentional when I say that. Ray was a unique soul and irreplaceable in so many respects, but he also represented what a person could become. Ray demonstrated how a business could be a powerful force for good when done right. He was the perfect example of doing well by doing good, a person who put purpose above profit but never sacrificed his competitive drive to succeed in the marketplace. His example can be replicated, and we need it to be. We need more Ray Andersons, and Ray's family, through the foundation created to carry on his work, is hoping to help create them.

Ray was diagnosed with cancer late in 2009, and while our family was devastated by the news at the time, I look back with gratitude on the twenty months we had with him before his passing. During that time, we celebrated his life as much as possible. Ray was not one for contemplating his mortality, though, so there were many questions that went unasked and unanswered. Questions like, *What should we do when you are gone?*

In reality I think that Ray avoided the topic because he did not want to rule from the grave. I respect him for that. Still, it meant that his daughters—my mother and aunt—were in for a surprise when they were called to the office of Ray's personal attorney about a month after Ray's death. For the first time they learned just how successful Ray had been, and that he was leaving the majority of his estate to the family foundation he had created.

"Mary Anne and I were both shocked and overwhelmed," my aunt, Harriet Langford, told me, "but really amazed that he left such a legacy for us to continue moving forward. Needless to say, we both felt a sense of panic over how big the shoes were for us to fill."

Harriet, along with my mother, Mary Anne Lanier, and their step-mother, Pat Anderson, were to be the three trustees of this nonprofit foundation. They had all known about the foundation's existence, as Ray and Pat had used it as a vehicle for their personal charitable giving for years. As a result of Ray's bequest, which he had always kept a secret, the foundation became rather more substantially endowed. There was just one hitch: Ray had not left any instructions about how to conduct the affairs of the foundation.

"I had no idea what we were going to do with the foundation, or where the money should go," my mother told me. "For a time we were angry with him because we didn't have the opportunity to work with him, to make a plan for the foundation, and to have a vision. I think that was part of his greater plan, though, and that he did that on purpose."

The foundation was a blank slate. Ray's wife and daughters were empowered to advance his legacy, but to do so in whatever way they saw fit. As Harriet told me, she came to see this as a gift rather than a curse. Ray had given them permission to go and figure out what charitable initiatives they felt had value.

That meeting with Ray's attorney in September 2011 was not the birth of the foundation in my eyes; nor was its formal creation two decades prior. We were still grieving as a family in the months that followed Ray's death, so setting a vision for the foundation was placed on hold. In particular, Mary Anne and Harriet were not quite ready to step into their father's shoes. They knew that they would need help to do so.

Which brings us back to Serenbe for a bright and sunny Tuesday in May 2012. The trustees had asked a group of influential people, each of whom knew and loved Ray, to join them in imagining a future for The Ray C. Anderson Foundation. Janine Benyus was there, as was John Picard, both members of Ray's EcoDreamTeam. Environmental philanthropist Julie Ann Wrigley attended, as did Laura Turner Seydel, daughter of Ted Turner and an environmental activist and philanthropist herself. Jo Ann Bachman, the foundation's first employee and previous assistant to Ray, joined the trustees at the meeting, which was facilitated by Mona Amodeo, who had received her PhD in organization development and change after studying Ray and Interface for her thesis. It was the foundation's first advisory board meeting.

"Nature was all around us, it was a beautiful day, and the birds were chirping," remembered Laura Turner Seydel. "And I think it was an emotional time, too, for the family to gather as Ray would have wanted."

Starting that day, Laura has been a guide to us in two important ways. As a trustee of the Turner Foundation and board chair of the Captain Planet

Foundation, she understood how to be an effective philanthropist. She also knew what it was like to be a part of a father's environmental legacy: "I learned so much from my dad when he started the Turner Foundation in 1990. It was such a great honor and privilege to be able to learn about the environmental movement from him. Ray and Ted have influenced millions of people around the planet and made such big, sweeping changes because of their vision, a vision combined with action."

I want to highlight two aspects of Laura's comment. First is the power of influence. From the beginning, Ray's influence has been one of the most valuable assets our foundation has had. Second is the importance of combining vision and action, which is as much the key to success for a philanthropist as it is for a businessperson. The foundation needed to first set the vision and then be active in executing it. After three days with the advisory board at Serenbe, that vision was set: So that society, business, and the environment might be harmonized for the benefit of generations yet to come, our foundation is committed to influencing and inspiring others to adopt the business and environmental values that Ray championed. That is what we mean by "creating more Ray Andersons."

How are we to live up to that high bar, though? After all, Ray was a radical industrialist. Moreover, there is a common trap into which foundations can slip—simply being a memorial to the founder. A grant here or there to keep a person's name alive rarely creates meaningful change. John Picard described this trap well. "I think people see foundations as second-tier, as if, 'Oh, he died, and now there's this little group of people trying to keep the legacy alive.'" Our task, then, was to become radical philanthropists in order to cultivate more radical industrialists.

Now several years into our own journey, we are seeing encouraging aspects of our work. In particular, one group of people has responded very well to our efforts to advance Ray's legacy, and it is a very important sector when you consider the economic transformation that we need in the coming decades: the business leaders of tomorrow. Fortunately, we have some remarkable partners in trying to instill Ray's values in young people.

Beril Toktay is one of them. In August 2012 she was beginning her eighth year teaching at Georgia Tech, but her first as a full professor. Other universities had recently tried to hire her away, but she was drawn to Georgia Tech's focus on technology and innovation and the growing expertise in its Scheller College of Business. She believed that Georgia Tech was well positioned to utilize those strengths to advance sustainability both at the institute and beyond. That summer she pitched a new center for sustainable business to the dean of the Scheller College, and while he was receptive, he also indicated that internal funding would have to be limited. Undeterred, Beril was ready to begin fund-raising to launch the center.

Around the same time, Mary Anne and Harriet decided to solicit proposals for funding from various universities, their first foray into grant making. Georgia Tech was one of those universities, and Beril threw her idea for this center into the ring. The three women sat down together in early 2013 to explore Beril's idea, which Mary Anne and Harriet had considered the boldest they had received. Ultimately, the trustees of our foundation made a multiyear commitment to create the Ray C. Anderson Center for Sustainable Business at Georgia Tech's Scheller College of Business, and Beril began working to fulfill her own vision. She wanted the center to create a virtuous loop among student, faculty, and industry interests. Faculty members would work on sustainability issues of interest to industry, and the center would be well equipped to stay abreast of the connections between industry and faculty research. Those connections would then create action-based, hands-on learning experiences for business students.

While this approach delivers value for multiple stakeholders, we are most excited for the impact it has on students. In nearly every meeting we have with the center's leadership, at least one student is invited to share their sustainability interests with us. They are amazingly brilliant and dripping with passion. In my mother's words, "These students get to start their whole careers with a sustainability understanding and then build upon it." It makes me wonder what Ray could have accomplished if he had received his spear in the chest before his sixtieth birthday!

These students will soon be joining the workforce, and my bet is that many of them will rapidly move into leadership positions. When they do, with their sustainability ethos well formulated by their experiences in college, they will work to change business and industry from within. They will become a meaningful part of the systems change and paradigm shift that we need.

They need not wait for the future to be impactful, however, as the Carbon Reduction Challenge, created by another Georgia Tech professor, Kim Cobb, demonstrates. Kim has been a climate scientist since earning her PhD in oceanography in 2002, focusing on global warming's impact on coral reefs. Her work has significantly contributed to the scientific community's understanding of El Niño events, and she has published dozens of peer-reviewed scientific papers. As accomplished as she may be in that field, however, I am most impressed by her innovation in the classroom. A few years after arriving at Georgia Tech in 2004, she approached her department chair about creating a new class. "I was chomping at the bit to design a new course for our department," she said, "one that would combine expertise across the entire campus. I pitched an energy/environment/society course to my chair and she said okay, but I didn't have a syllabus and I had very little idea of what I was doing. I couldn't test the students on climate science because it was a small, three-credit lecture, and I didn't want them gnawing on a research problem and not get anywhere. So I decided to challenge them to do something in the real world."

In the Carbon Reduction Challenge, students form teams and design projects for external partners that will measurably reduce the concentration of greenhouse gases in the atmosphere. They are graded in part on how many pounds of carbon dioxide equivalent they are *actually* able to remove, with a minimum goal of a ten-thousand-pound reduction. Kim works with the students to find the external partners, often businesses in the Atlanta area, who can execute on the projects designed by the students. Her teams calculate the carbon reduction, upfront financial costs, and lifetime financial savings of their ideas, and then make the pitch to the external partners to adopt and execute the program.

Over the years students have developed some amazing projects. One of Kim's favorites is a team of five women who wanted to put a cool roof on a building at Georgia Tech. Kim was originally skeptical that they could shepherd the project to completion, but the women made a compelling pitch to Georgia Tech's facilities department. They projected that the institute could save $5,245 annually in energy costs with an upfront cost of $46,150, resulting in an anticipated payback period of 8.8 years. With those numbers, the students convinced the facilities department to get behind the project, which was endorsed by Georgia Tech's administration the day before the class deadline. Over the twenty-year warranty period for the roof, estimated carbon savings amounted to 459 metric tons, which easily won the challenge for these women that year.

Impressed by Kim's track record, the NextGen Committee of our foundation (consisting of Ray's five grandchildren and our spouses) decided to make a grant to Georgia Tech to expand the concept of the class. With our support, and in partnership with Beril Toktay and the Ray C. Anderson Center for Sustainable Business, Kim created a Carbon Reduction Challenge for Georgia Tech students during their 2017 summer internships. In addition to working on the tasks for which their employers hired them, the students who chose to participate would design projects that would reduce their employers' carbon footprints.

I was stunned by the breadth and creativity of every student project that summer, but the team that won the challenge stood out for the elegance and simplicity of their project. Six students were interning for a large national bank, and in their hunt for carbon savings, they focused on the bank's travel policy for employees. When employees would book travel for work and need to rent a vehicle, they were given the choice of what size vehicle they would like. If those employees did not indicate a preference, the travel agent would default to booking an intermediate-sized vehicle. By simply changing the default rental car option to a fuel-efficient economy vehicle, these students projected that the bank could save $40,000 annually with accompanying carbon savings of forty-five metric tons of carbon dioxide equivalent per year. That is some low-hanging carbon and financial fruit! Even better, these students have seen just how impactful they can be. They are already making a tremendous difference, and they will only continue to do so when they leave Georgia Tech.

Our efforts to turn the business leaders of tomorrow into more Ray Andersons also extends beyond the walls of academia. Businesses that have not yet been dreamed of will have a transformative influence in the decades to come, so we believe that influencing aspiring entrepreneurs can also accomplish our goal. In this respect, we have returned to a source of inspiration that Ray himself loved—biomimicry.

Since its founding in 2006 by Janine Benyus, the Biomimicry Institute has sought to show the world that nature really is the best teacher. When biomimicry has become a ubiquitous design discipline rather than a niche one, you will know that they have succeeded. To further that goal, the institute began a student design challenge in 2008, allowing teams from around the world to develop biomimetic products and solutions to thematic challenges, such as water scarcity or energy use.

Response to the challenge was remarkable in those early years. Teams entered from Iran, Nicaragua, Sweden, Egypt, and many other countries, in addition to teams from the United States. Many of their biomimetic designs were brilliant, but all were limited by the design challenge's conclusion point. Because of a lack of resources, the institute could only offer a $5,000 prize to one winning team, with no continuing support to teams who might want to then commercialize their designs. The designs did not have any realistic way to move forward.

In the spring of 2014, Harriet, Mary Anne, and I sat down with Janine and Beth Rattner, executive director of the Biomimicry Institute, and talked about what it would take to allow these aspiring biomimics to take the next step and how to expand the challenge beyond students to anyone with the courage to become an entrepreneur. What resulted from those conversations is an initiative called the Biomimicry Global Design Challenge.

As with the previous student design challenge, teams from around the world are encouraged to develop concepts for a biomimetic product or service. The participants with the best ideas, along with other applicants who were already beyond the design phase, are then invited to move forward to the Biomimicry Launchpad, a business accelerator program that supports the teams with mentorship, training, and prototyping support. At the conclusion of the Launchpad, the teams compete for the annual $100,000 Ray of Hope Prize, provided by our foundation. They are judged on the quality of their biomimicry, their commercialization potential, and the positive social and environmental impacts of their designs.

The first Ray of Hope Prize was awarded in October 2016 at the Bioneers annual conference in San Rafael, California. The winning team that year had developed an application for degraded soil that could, they hoped, restore the soil to health in one growing season. The second year, the winning team developed a passive water capture and storage system that could support agriculture in arid climates and in dense cities. Importantly, though, even the teams who do not win the grand prize are encouraged to continue, and some of the most successful teams in the long run have been those that did not win a grand prize.

If Ray were still here today, he would be so impressed by these budding entrepreneurs, even though he would probably dismiss the idea of creating more Ray Andersons; he would not want the goal to be so focused on him. Nonetheless, his name and example are valuable to their work. "Honestly, having Ray Anderson's name attached is important because of Ray's story," noted Beth Rattner. "It matters that he was an industrialist, a former 'pillager,' and that he experienced a transformation that made enormous ripple effects of change."

I have seen firsthand the impact that we are making in the lives of young students and entrepreneurs, so I know that our foundation's goal is a good one. I have also seen the ways in which we have unexpectedly succeeded. It turns out that we are creating Ray Andersons in ways that we never imagined, and one person in particular stands out in that regard: his daughter Harriet.

Early in 2014 Amy Lukken and Wendell Hadden, both longtime members of the Interface family, approached Harriet with the idea of naming a road after Ray. They suggested the industrial park in LaGrange, Georgia, which is home to Interface's North American manufacturing operations, but Harriet wanted to think bigger. Interface has operations both in LaGrange and West Point, Georgia, Ray's birthplace. Those two towns are connected by Interstate 85 and only about eighteen miles apart, so Harriet and her husband, Phil Langford, got in touch with their local state representative, Randy Nix. With Representative Nix's help, and just before the close of the legislative session in March 2014, the state legislature formally renamed a portion of Interstate 85 in West Georgia the "Ray C. Anderson Memorial Highway."

As a family, we were honored. Harriet and Phil were particularly pleased, as proud residents of LaGrange themselves. In the weeks following the announcement, however, Harriet began to feel unsettled. Interstate 85, like all interstates, is a dirty highway. She had just put the name of the greenest industrialist on a dirty highway.

At the time, Harriet was a member of the board of directors of the Georgia Conservancy. She had developed a friendship with Allie Kelly, then the conservancy's senior vice president, so she called Allie with a particular question: "Has anyone tried to make a highway sustainable?"

Apparently not. While certain locales were testing isolated environmentally friendly roadway technologies like wildlife bridges or solar farms in the right-of-way, Harriet and Allie could not find any road in the world that was holistically designed to be sustainable, much less regenerative. Some people might let the question go at that point. Harriet could not. "I couldn't rest, and I felt a sense of urgency," she said. "I put his name on a dirty highway! It was my own spear-in-the-chest moment, my own epiphany that we could do better. If Ray could do it in industry, surely we can do it in transportation."

Harriet was onto something. If we are to see a broad shift of business toward sustainability, we will need to see other systems that are integrally connected to it shift in concert. Transportation is one such system, and American highways in particular have not been innovated in a meaningful way since the development of our Interstate Highway System in the 1950s. Perhaps Ray Anderson's approach toward industrial ecology would be successful in transportation. He sought to build the example of a sustainable industrial manufacturing company that would prove what is possible and reveal sustainability to be a better way. Could these eighteen miles of interstate be the same thing for transportation?

We decided to create and fund a new nonprofit, now called The Ray, to house the project, and Harriet convinced Allie to join as its executive director. In partnership with the Georgia Conservancy, Georgia Tech, Innovia Technology, and ICON, Harriet and Allie worked to understand what a sustainable highway would look like and how this new organization could best lead its creation. To the first point, they learned that a sustainable highway would filter stormwater, remove pollutants, generate electricity, be made of renewable materials, conserve wildlife, have zero human deaths, sequester carbon, support the development of emergent vehicle technology, support local communities, and change attitudes about environmental stewardship. Quite a list! To the second point, they learned that The Ray should provide thought leadership and facilitate partnership between the governmental agencies that oversee our roads and the businesses that are innovating new roadway technologies.

Four years into the project, deployed technologies include a photovoltaic electric vehicle charging station sponsored by Kia Motors, a pilot test of drive-on solar panels called Wattway, bioswales adjacent to the highway to filter pollutants, a tire-safety monitoring station called WheelRight, a pollinator garden, and planting of perennial wheat in the right-of-way. Just around the corner, The Ray anticipates a large solar farm in the right-of-way and drone monitoring of the corridor, allowing for better measurement and tracking of the highway's conditions. The Ray is even pioneering its own technology: smart studs in the striped lines that can monitor the road and communicate traffic conditions to drivers, thereby increasing human safety and perhaps enhancing the capabilities of autonomous and connected vehicle technologies currently in development.

None of this could have happened without the Georgia Department of Transportation (GDOT), and fortunately it has been a willing and engaged partner in bringing this highway of the future to life. GDOT realizes that many of these roadway innovations offer economic opportunities in addition to putting Georgia on the map. Our highways and the adjacent rights-of-way have inherent value and revenue generation potential that has been untapped for decades, and GDOT is taking the first steps with The Ray to unlock that value. We firmly believe that GDOT can do well by doing good, just as Interface has.

For Allie, the greatest success of The Ray so far is that it is "a functioning living laboratory, and within a pretty short period of time. Had we known how fast it could all happen at the beginning, we might have thought twice. It's hard to be doers."

Allie is probably right, but I selfishly see my aunt's personal transformation as the greatest success of The Ray. If she was before committed to advancing Ray's legacy, she is now equally committed to living it. The Ray will be her life's work and her own legacy, and she has been both inspired and empowered by her father. She is the best example I can give of why we seek to create more Ray Andersons.

She is also an example of what Ray emphasized in his last several years of public speaking. Women, he said, were the great hope of the environmental movement; empowering them to become leaders would bring new thinking, and usually better thinking, to the challenges that we need to solve. I believe that our foundation is showing, even if just in a small way, how right Ray was. We did not plan it this way, but it turns out that every major initiative that we support is led by women.

Nearly a decade has passed since Ray's death, but it seems as if the world has seen more than a decade of change. The climb up Mount Sustainability is taking ever more twists and turns. Crevasses are growing wider as we experience the pressures of exponential growth: more people, more resource use, more climate impacts. We are rushing toward natural limits. In the process, though, many of us have refocused our lenses. Where we used to see a need for business-level change, we now see an additional need for system-level change: Our focus on encouraging prototypical companies of the twenty-first century has widened to encouraging the prototypical economy of the twenty-first century.

Despite the massive scale of our challenges, take solace: We will create this change together, and the weight of the world's problems need not fall on one person. Worry less about the fate of humanity and more about the incredible opportunity we have to create a better world for Tomorrow's Child. The generations to come will be our judges, as elegantly put by Laura Turner Seydel: "The measure of our legacy will be the moment on our deathbed when we look our children and grandchildren in the eye and say, 'I've done everything that I can possibly do for you and your children.' And for them to look us in the eye and say, 'Yes, I know you have.'"

Ray did all that he could possibly do. Will you?

Ray's Acknowledgments

From the Original *Mid-Course Correction*

I begin by acknowledging my debt to the past, personified by my parents. I think they would have been pleased by reading this and would write off their portion of that debt.

This book, drawn from thoughts developed in the process of preparing and giving some two hundred speeches, would not exist without the infinite patience of the people who have transcribed my inadequate literary attempts through uncounted iterations. Janet Amundsen, my administrative assistant, slugged through the drafting and redrafting of all my earliest speeches upon which the bulk of the book is based; Kaye Gordy filled in during Janet's leave of absence; and, most recently, Cindy Stringer labored through each "fine-tuning" redraft. My sincerest thanks go to each of them.

I thank Jim Hartzfeld and Jennifer DuBose for their assistance in working out the schematics used in chapter 5 and for compiling the list of ideas in the appendix.

I acknowledge my wife, Pat, for her loving endurance of my many nights away from home as I have crisscrossed North America and two oceans to deliver the speeches that have consumed my life for the last three years.

I acknowledge the special contribution of Paul Hawken, Melissa Gildersleeve, Joyce LaValle, John Picard, and Gordon Whitener, which is revealed near the end of chapter 7.

I also acknowledge the contributions of two new friends who have come into my life in connection with my membership on the President's Council on Sustainable Development: Jonathan Lash and Timothy Wirth. Tim's comments after reading an early draft were very encouraging, especially in light of his position at the time as undersecretary of state for global affairs and, currently, as president of Ted Turner's United Nations Foundation. Jonathan's encouragement, also after reading an early draft, to write a "bigger book" influenced me to expand the part about myself that also deals with the founding of my company, Interface, Inc. Jonathan, I am afraid it is still a little book, but thank you for pushing me.

I acknowledge with special thanks the suggestions offered by Mike Bertolucci, Bill Browning, Paul Hawken, Amory Lovins, Fred Krupp, and Peter Raven that helped me tighten up the factual accuracy of some of the technical discussions, and the help early on from Phyllis Mueller in editing preliminary drafts and, at the end, for cleaning up after me.

In many ways most important of all, I acknowledge the manifold ways that the people of Interface, led by a group called the Peregrinzillas (they know who they are), are actually doing the things to change our company—the changes that give the words of this book whatever credibility they possess.

I conclude by acknowledging my debt to the future, personified by my grandchildren, and their children, and theirs. I hope that their world will be a bit better than it otherwise would have been.

—RAY C. ANDERSON
July 1998
Atlanta, Georgia

ACKNOWLEDGMENTS

John's Acknowledgments

W hen Chantel and I exchanged our marriage vows, she accepted me "for better" and "for worse," but not "for a book." I am truly lucky to have a wife who would so willingly walk this path with me. Her patience with this process has been exceeded only by her support and love for me and for our young children. Thank you, Chantel. I could not have written this without you. Neither would I have wanted to.

To Mary Anne Lanier, Harriet Langford, Jaime Lanier, and Phil Langford, thank you for believing that I was the right person to continue telling Ray's story. Working with you to advance his legacy has been one of the highest honors of my life. You should be proud of what you have accomplished in leading the Ray C. Anderson Foundation, and I hope you are just as proud of this book.

To Jay Lanier, Whitney Lanier, Patrick Lanier, Stephanie Lanier, Melissa Heflin, McCall Langford, and my wife (once more), thank you for your support of me and for your enthusiasm and passion for our foundation's NextGen Committee. I know that Ray smiles from the heavens when he sees us gather together. I look forward to many more decades of learning from you all.

To Valerie Bennett and Lori Blank, thank you for shouldering the load at the foundation while I immersed myself in writing. Our work was in the best of hands during my "absence," and I could not ask for a better staff. You are very much a part of this family.

To Lisa Lilienthal, thank you for the idea for this book! You were right—it had another life to live. Your guidance and hard work throughout this process have been invaluable. My gratitude also goes to your colleagues at ICON who were working behind the scenes.

To Margo Baldwin and Joni Praded at Chelsea Green Publishing, thank you for your "yes"! Given how much you loved Ray, we simply could not have had a better publisher and editor. Thank you as well for taking a chance on this first-time author.

To Alan Anderson, Janine Benyus, Bob Fox, Paul Hawken, Steve Nygren, John Picard, Laura Turner Seydel, John Wells, and Julie Wrigley, I am eternally grateful for the lessons you have taught me as members of our foundation's advisory board. That I get to count you as friends as well as teachers is an honor of which I am not worthy.

To everyone whom I interviewed for this book, thank you for our enriching conversations. More important, thank you for everything that you do to advance sustainability in your own ways. You are all inspirations to me.

To the grantees of the Ray C. Anderson Foundation, thank you for your partnerships. Effective philanthropy requires humility, and we at the foundation know that your efforts are what actually create change in the world. We feel privileged to be your supporters.

To my dear friends at Taproom Coffee and East Pole Coffee Co. in Atlanta, thank you for being my "homes away from home." Nearly every word that I wrote in this book was written in your fine establishments. The coffee was amazing, but the conversations were better.

To my other Atlanta friends, thank you for your simple presence in my life. I enjoy every meal, drink, and celebration that we share, and I apologize for having shared so few of those the past several months. I am fortunate to have you as my "tribe."

Finally, I echo my grandfather's own words in acknowledging the people of Interface. What you have accomplished is remarkable. What you will accomplish next will be even more so. Thank you, and keep climbing, my friends!

—JOHN A. LANIER
August 2018
Atlanta, Georgia

APPENDIX

Interface's Original Sustainability Checklist

Circa 1998

PEOPLE:

Customers

- Provide honest information about the known environmental impacts of your company and product
- Invite customers to audit and critique your efforts
- Share your understanding of environmental issues and natural systems with customers

Employees

Culture

- Create an atmosphere that encourages employees to question the status quo and take risks
- Create an environment that encourages lifelong learning
- Engage the creativity of all employees and associates

Understanding

- Educate all employees on the corporate sustainability vision
- Educate all employees on basic environmental principles and the workings of natural systems
- Create mechanism for employees to share knowledge of best practices
- Bring in experts to address and challenge employees
- Create newsletters to report sustainability projects and challenges, including information that is not specific to the company
- Provide access to information that can help employees in their private lives; for example, sponsor seminars on ways to save energy at home
- Have a "dumpster diving" activity to understand the makeup of your waste stream

- Use experiential learning techniques to explain complex concepts
- Hold a seminar to explain the dos and don'ts of your recycling program

Involvement
- Ask employees to give input into improving environmental impacts of their jobs
- Ask employees if there are easy or low-cost things that the company could do to make their jobs more pleasant and them more productive
- Involve employees in decision making when it affects them
- Always listen to what employees have to say about issues that affect them
- Respect the knowledge and intelligence of all employees
- Create work group teams to eliminate waste in their work areas

Suppliers
- Share your corporate vision and internal framework for sustainability with suppliers
- Involve suppliers in educational opportunities to learn more about sustainability

Community
Environmental Organizations and Government Programs
- Partner with environmental organizations that work on issues important to your corporate philosophy
- Commit a percentage of profits to environmental research
- Participate in voluntary government programs with the Environmental Protection Agency, such as Green Lights, Energy Star Buildings, and Climate Wise

Networking
- Contact other companies with a similar vision; share ideas
- Work with local universities to find the latest environmental technologies and understanding
- Work with universities in joint research projects
- Talk with global experts
- Search for good practices and ideas outside your company
- Share your accomplishments with others and multiply good practices through them
- Invest time and resources in organizations committed to environmental progress or sustainable development

The Public
- Develop auditing mechanisms open to public disclosure
- Make public statements in support of sustainability principles and public disclosure such as the CERES (Coalition for Environmentally Responsible Economies) Principles
- Sponsor community forums about local environmental issues
- Choose community projects to support with time and money
- Open facilities to local schoolchildren to learn about sustainability and career opportunities

Management

Corporate Strategy
- Establish top management commitment to long-term environmental strategy
- Establish corporate and divisional sustainability vision statements
- Ask for volunteers to serve as local environmental coordinators
- Establish local Green Teams to implement ideas
- Gain certification in third-party-assessed environmental management systems such as ISO 14001 or BS 7750
- Create a process of managing all aspects of environmental stewardship
- Develop well-defined corporate values, goals, decision making, and response mechanisms
- Evaluate product and service offerings for fit with a sustainable society

Metrics
- Measure all material and energy flows in physical and monetary units
- Develop a managerial "Full Cost Accounting" system
- Audit management systems and disposal practices
- Measure material and energy flows per unit of output to adjust for changes in production levels
- Create internal "green taxes" to highlight most profitable enterprise from a total cost perspective

Incentive Plans
- Give rewards to individuals or teams with the best sustainability project
- Tie monetary compensation to achieving well-defined environmental goals
- Recognize outstanding commitment and progress toward sustainability

Keeping the Enthusiasm
- Set reasonable goals and always celebrate your accomplishments

- Learn through playing games
- Develop a sense of competition and pride

- Bring in college interns to research special projects for a fresh perspective
- Volunteer for a local hands-on project as a corporate team where the results of your labor are almost immediate; for example, plant a garden of native plants

PRODUCT:

Design

- Redesign products to use less raw materials while delivering the same or greater value
- Replace nonrenewable materials with more sustainable materials, such as:

  - Organic materials: products of nature such as wood, cotton, hemp, flax, vegetable oils, et cetera
  - Organically grown and sustainable harvested materials, such as organic cotton and produce, certified wood products, et cetera
  - Locally produced and abundant materials
  - Recycled and reclaimed post-consumer or post-industrial waste materials
  - Materials consuming lower embodied energy

- Eliminate the use of hazardous chemicals
- Design products to minimize consumption of energy and auxiliary materials in use
- Design products to last longer; make products more durable
- Design products to be repaired or selectively replaced when only a portion wears out
- Develop products out of easily separated components, or out of only one material, to facilitate recycling
- Consider the entire life cycle of a product, including how it will be recovered and made into another useful product

Packaging

- Design out all product packaging, like a "taco shell" (the package is part of the product)
- Develop returnable packaging
- Deliver products in bulk
- Develop reusable packaging for work-in-process materials
- Use recycled materials
- Design packaging to be more easily recycled

- Design packaging to be safe or biodegradable if accidentally released into the environment

Manufacturing

Energy

DEMAND

- Reclaim waste heat from processes, furnaces, air compressors, and boilers
- Systematically review all electric motor systems to minimize installed horsepower and maximize motor efficiency
- Design pumping systems with big pipes and small motors
- Design pumping systems by laying out pipes first (to minimize distance and elbows), then motors and other equipment
- Lay out plants to minimize distance materials travel
- Research product formulations to reduce process temperature requirements
- Minimize the number of times materials are heated or cooled
- Install multiple small motors to handle varying volumes rather than one big motor
- Design systems for expected operating conditions rather than maximum expected capacity
- Stage plant flows and energy peaks to maximize efficiency
- Use computer modeling techniques to minimize energy usage
- Research Energy Miser technology on motors
- Install power submeters on all processes to continuously monitor efficiencies
- Install automatic switches to turn off equipment at a determined time of inactivity

SOURCE

- Research and adopt alternative energy sources consistent with local surroundings, such as hydroelectric, biofuel, solar, wind power, et cetera
- Negotiate green energy contracts with utilities
- Research soft starting / control motor technologies
- Research energy storage technologies such as flywheels

Material

- Adopt a zero waste mentality; design processes to create no waste or scrap
- Adopt a zero defect mentality; most material defects become waste
- Eliminate all smokestacks, effluent pipes, and hazardous waste
- Adopt high-efficiency planning and scheduling practices to minimize waste

- Network with other companies to find waste streams that can become inputs for other processes
- Buy raw materials in bulk to minimize packaging
- Carefully segregate waste materials for reuse or recycling
- Develop processes to utilize internal scrap materials
- Develop quick-stop technology to minimize waste created by off-quality processes
- Take corrective action on quality problems as far upstream as possible to minimize waste
- Closely measure all material streams to monitor material efficiency

Marketing

- Commit to taking back your products at the end of their lives
- Rent only the service component of your products, such as warmth and light, rather than sell the product
- Be conscious about the extent and strategy of external communications to avoid greenwash

Purchasing

Work with Suppliers

- Share your corporate purchasing policy with all suppliers
- Press suppliers to follow and document sustainable practices, and favor those that do
- Press suppliers to take back packaging or not deliver it with the product
- Buy services, not products
- Encourage suppliers to report their environmental impacts in your terms
- Encourage suppliers to develop and offer products with a smaller environmental footprint
- Ask for information about the environmental policy of the corporation and information about the specific products you buy from suppliers
- Include the waste and embodied energy used to produce raw materials purchased from suppliers in your environmental footprint analysis

Buy Sustainably

- Establish a "Buy Sustainably" policy stating the corporate goals on specific items when possible
- Circulate a list of recycled or environmentally friendly products to purchasing staff
- Set out clear guidelines to follow
- Support training for purchasing agents to understand the issues

- Create an internal purchasing agent team focused on identifying appropriate products
- Develop environmentally responsible methods of reconditioning used products
- Share surpluses with other offices by publishing a regular list
- Implement high-efficiency planning and scheduling practices to minimize waste

PLACE:
Facility
Design

- Increase insulation in walls and doors
- Use double-paned or super windows
- Use high-efficiency glazing or films on windows
- Use shades, deflectors, and light shelves to reduce summer sun
- Design HVAC and utility systems for maximum long-term flexibility and efficiency, such as under floor delivery, personal control
- Maximize use of natural ventilation heating and cooling
- Specify finishes and materials with low VOCs and that control the growth of microbial contamination
- Install fast-acting doors in factory and warehouse exits to minimize time the door is open to the outside
- Design a minimum of impermeable surfaces to minimize stormwater runoff
- Give priority to pedestrians, mass transit riders, and cyclists instead of automobile drivers
- Design with the natural flows of the site in mind
- Provide safe areas to secure bicycles
- Use minimal finishes, such as paints and coatings
- Use low-embodied-energy, locally abundant building materials
- Rehabilitate existing buildings rather than tear them down
- Use salvaged or refurbished materials
- Locate near existing infrastructure
- Separate and recycle construction waste

Operations

ENERGY

- Conduct an energy audit with the help of local utilities
- Replace old boilers with new high-efficiency units
- Install properly sized variable-speed motors/fans

- Install heat exchangers on building exhaust ducts
- Preheat boiler feedwater with waste heat
- Use excess plant heat to heat offices
- Install programmable thermostats
- Regularly maintain all HVAC systems
- Check fan speeds and efficiencies on HVAC systems
- Regularly replace filters and clean ductwork
- Replace CFC in cooling systems with non-ozone-depleting refrigerants
- Install variable air diffusers
- Plant trees to shade eastern, western, and southern windows and air conditioners

Lighting
- Redesign lighting to fit work processes, resulting in productivity improvements
- Install infrared motion detectors for automatic lighting control
- Replace incandescent lighting with compact fluorescent lighting
- Retrofit existing lighting with high-efficiency fluorescent or metal halide bulbs, electronic ballasts, and reflectors
- Reduce the use of high bay lighting
- Maximize the use of natural daylight

Water
- Reuse water whenever possible
- Reuse boiler water
- Treat and reuse dye water
- Develop closed loops whenever possible
- Conduct water-use audits, looking for leaks and waste
- Install low-flow fixtures in restrooms and kitchen areas

Office

PAPER
- Use recycled paper with a high percentage of post-consumer content
- Use chlorine-free paper, if available
- Use paper envelopes without windows and avoid Tyvek envelopes, so envelopes can be recycled
- Place collection containers at every work station to recycle used paper
- Reduce or eliminate paperwork and numbers of copies
- Scrutinize distribution lists
- Make copies only on request; otherwise, route material

- Maximize use of bulletin boards
- Set up copiers so that double-sided copying is the norm
- Route magazines instead of getting separate copies
- Keep paper that is still good on one side (GOOS paper, Good On One Side) and make scratch pads out of it
- Communicate via email when possible, and don't print your email messages
- Eliminate cover sheets on faxes

ELECTRONICS

- Purchase only energy-saving electronic equipment—look for the EPA's Energy Star label
- Turn off computer monitors when not in use
- Turn off your computer when you go to lunch and overnight
- Use a projector instead of printing overheads for presentations
- Send used overheads back to 3M to be recycled
- Send exhausted ink jet cartridges back to their manufacturer for recycling
- Use refillable ink jet cartridges
- Lease the service of high-end electronics instead of buying them (then they can be returned to the provider when you decide to upgrade instead of being disposed of)
- Buy copiers, printers, and fax machines that use refurbished parts and toner cartridges

Maintenance

- Invest in high-quality maintenance to extend the life and maximize the efficiency of systems
- Use only nontoxic cleaning compounds
- Maximize the use of all-purpose cleaners to reduce the number of chemicals used and to minimize potential danger of mixing
- Buy cleaners in concentrated forms that can be mixed at different strengths for different purposes, reducing packaging and transportation
- Use washable mugs, glasses, plates, and utensils
- Use bulk product dispensers for beverages, condiments, et cetera
- Provide convenient and easy-to-understand recycling centers for common waste products
- Measure all solid waste streams

Landscape

- Leave as much habitat and vegetation as possible undisturbed by construction
- Landscape to promote biological diversity

- Design to minimize impact on the local environment
- Compost organic matter
- Mulch lawn clippings
- Put up bird boxes and start an employee-run nest box monitoring program
- Plant a butterfly garden near an area that employees use often
- Join the Wildlife Habitat Council
- Start an employee vegetable garden
- Create a series of nature trails for employees and their families or even for the whole community
- Xeriscape by using plants adapted to local rainfall conditions
- Use gray water to water the landscaping
- Highlight native plants that are adapted to the local environment and do not require a lot of maintenance
- Employ integrated pest management to minimize the use of chemical pesticides
- Install stormwater retention ponds to minimize volume and temperature spikes on local waterways from rain showers
- Create bird sanctuaries in migration paths

Transportation

Product

- Ship by rail whenever possible
- Reduce the weight of products to consume less energy in transport
- Favor locally produced products
- Create transportation consortiums to maximize the loading of trucks with other local businesses
- Pelletize waste materials such as fiber to minimize transportation energy
- Locate facilities to minimize shipping distances to major market centers

People

- Offset employee travel and product transportation with tree planting through organizations such as Trees for Travel
- Reduce number of trips by consolidating business or through better planning
- Buy alternative fuel vehicles
- Allow employees to telecommute or work alternative hours
- Offer rebates to employees who use alternative transportation and do not consume parking space
- Offer public transportation passes to employees
- Encourage videoconferencing

Notes

Chapter 1. The Next Industrial Revolution

1. Rachel Carson, *Silent Spring* (Boston: Houghton Mifflin, 1962).
2. Lester C. Thurow, "Brains Power Business Growth," *USA Today*, August 18, 1997, 13A.
3. Paul Hawken, *The Ecology of Commerce: A Declaration of Sustainability* (New York: Harper Business, 1994).
4. Walter R. Stahel, "The Product-Life Factor," Houston Area Research Center, 1982, http://www.product-life.org/en /major-publications/the-product-life-factor
5. Karl-Henrik Robèrt, "Educating a Nation: The Natural Step," *In Context* 28 (Spring 1991): 10.
6. Paul Ehrlich, *The Population Bomb* (Cutchogue, NY: Buccaneer Books, 1997).

Chapter 2. A Spear in the Chest and Subsequent Events

1. Daniel Quinn, *Ishmael* (New York: Bantam Books, 1993).
2. Daniel Quinn, *The Story of B: An Adventure of the Mind and Spirit* (New York: Bantam Books, 1997).
3. Daniel Quinn, *My Ishmael: A Sequel* (New York: Bantam Books, 1997).
4. Al Gore, *Earth in the Balance: Ecology and the Human Spirit* (New York: Plume, 1993).
5. Donella H. Meadows et al., *Beyond the Limits: Confronting Global Collapse, Envisioning a Sustainable Future* (White River Junction, VT: Chelsea Green Publishing, 1993).
6. Lester R. Brown et al., *Vital Signs 1997: The Trends That Are Shaping Our Future* (New York: W. W. Norton, 1997).
7. Lester R. Brown et al., *State of the World 1998: A Worldwatch Institute Report on Progress Toward a Sustainable Society* (New York: W. W. Norton, 1998).
8. Joseph J. Romm, *Lean and Clean Management: How to Boost Profits and Productivity by Reducing Pollution* (New York: Kodansha, 1994).
9. Herman Daly and John Cobb, *For the Common Good: Redirecting the Economy Toward Community, the Environment, and a Sustainable Future* (Boston: Beacon Press, 1994).
10. Joseph L. Bast et al., *Eco-Sanity: A Common-Sense Guide to Environmentalism* (Lanham, MD: Madison Books, 1994).
11. Ronald Bailey, ed., *The True State of the Planet: Ten of the World's Premier Environmental Researches in a Major Challenge to the Environmental Movement* (New York: Free Press, 1995).

Chapter 4. A Mountain to Climb

1. Karl-Henrik Robèrt et al., "A Compass for Sustainable Development," *The Natural Step News* 1 (Winter 1996): 3.
2. Donella H. Meadows, "Places to Intervene in a System (in Increasing Order of Effectiveness)," *Whole Earth* (Winter 1997): 78.
3. Abraham Maslow, *Motivation and Personality* (Boston: Addison-Wesley Publishing, 1987).
4. Immanuel Kant, *The Critique of Practical Reason* (Germany: 1788).

Chapter 5. The Prototypical Company of the Twenty-First Century

1. Fritjof Capra, *The Turning Point: Science, Society, and the Rising Culture* (New York: Bantam Books, 1988).
2. Peter Russell, *The Global Brain Awakens: Our Next Evolutionary Leap* (Palo Alto, CA: Global Brain, 1995).
3. Brian Swimme, *The Universe Is a Green Dragon: A Cosmic Creation Story* (Rochester, VT: Bear, 1988).

Chapter 7. To Love All the Children

1. Glenn C. Thomas, "Tomorrow's Child," copyright 1996. All rights reserved. Reprinted with permission.
2. The President's Council on Sustainable Development, *Sustainable America: A New Consensus for Prosperity, Opportunity, and a Healthy Environment for the Future* (Washington, DC: The President's Council on Sustainable Development, 1996).

Chapter 8. Nearing the Summit

1. "Global Greenhouse Gas Emissions Data," Environmental Protection Agency, accessed August 28, 2018, https://www.epa.gov/ghgemissions/global-greenhouse-gas -emissions-data.
2. "Mission Zero: Measuring Our Progress," Interface, accessed August 28, 2018, http://www .interface.com/US/en-US/campaign/climate-take-back/Sustainability-Progress-en_US.
3. Ray C. Anderson, "The Business Logic of Sustainability," filmed May 2009 at TED2009, TED video, 15:45, https://www.ted.com/talks/ray_anderson_on_the_business_logic _of_sustainability.

Chapter 9. All or Nothing

1. Mark Mykleby et al., *The New Grand Strategy: Restoring America's Prosperity, Security, and Sustainability in the 21st Century* (New York: St. Martin's Press, 2016), 138.
2. Janine Benyus, *Biomimicry: Innovation Inspired by Nature* (New York: HarperCollins, 1997), 7.
3. Daniel Esty and Andrew Winston, *Green to Gold: How Smart Companies Use Environmental Strategy to Innovate, Create Value, and Build Competitive Advantage* (Hoboken, NJ: John Wiley & Sons, 2009), 217.

4. Jeffrey Hollender and Bill Breen, *The Responsibility Revolution: How the Next Generation of Businesses Will Win* (San Francisco: Jossey-Bass, 2010), 67.

5. Hollender and Breen, *The Responsibility Revolution*, 19.

6. Esty and Winston, *Green to Gold*, 129.

Chapter 10. The Prototypical Economy of the Twenty-First Century

1. Paul Hawken et al., *Natural Capitalism: The Next Industrial Revolution* (New York: Little, Brown, 1999), 4.

2. Hawken et al., *Natural Capitalism*, 153.

3. Hawken et al., *Natural Capitalism*, 5.

4. Donella H. Meadows et al., *Limits to Growth: The 30-Year Update* (White River Junction, VT: Chelsea Green Publishing, 2004).

5. Andrew Winston, *The Big Pivot: Radically Practical Strategies for a Hotter, Scarcer, and More Open World* (Boston: Harvard Business Review, 2014), 70.

6. Kate Raworth, *Doughnut Economics: 7 Ways to Think Like a 21st Century Economist* (White River Junction, VT: Chelsea Green Publishing, 2017), 46–47.

7. Raworth, *Doughnut Economics*, 44.

8. Raworth, *Doughnut Economics*, 45.

9. Raworth, *Doughnut Economics*, 88.

10. Raworth, *Doughnut Economics*, 209.

Chapter 11. Bending the Line

1. Stahel, "The Product-Life Factor."

2. Winston, *The Big Pivot*, 39.

3. Yvon Chouinard, *Let My People Go Surfing: The Education of a Reluctant Businessman* (New York: Penguin Books, 2016), 83.

4. Chouinard, *Let My People Go Surfing*, 83.

5. Rachel Botsman and Roo Rogers, *What's Mine Is Yours: How Collaborative Consumption Is Changing the Way We Live* (London: HarperCollins, 2011), 71.

6. Botsman and Rogers, *What's Mine Is Yours*, 75.

7. Botsman and Rogers, *What's Mine Is Yours*, 83.

Chapter 12. A New Perspective on Our Climate

1. Paul Hawken, ed., *Drawdown: The Most Comprehensive Plan Ever Proposed to Reverse Global Warming* (New York: Penguin Books, 2017), 222.

2. Hawken, *Drawdown*, 223.

Index

About the Authors

Ray C. Anderson was founder and chairman of Interface, Inc., one of the world's leading carpet and flooring producers. His story is now legend: Ray had a "spear in the chest" epiphany when he first read Paul Hawken's *The Ecology of Commerce*, inspiring him to revolutionize his business in pursuit of environmental sustainability. In doing so Ray proved that business can indeed "do well by doing good." His Georgia-based  company has been ranked number one in a GlobeScan survey of sustainability experts, and it has continued to be an environmental leader even after Ray's death in 2011. Ray authored the 1998 classic *Mid-Course Correction*, which chronicled his epiphany, as well as a later book, *Confessions of a Radical Industrialist*. He became an unlikely screen hero in the 2003 Canadian documentary *The Corporation*, and was named one of *Time* magazine's Heroes of the Environment in 2007. He served as co-chair of the President's Council on Sustainable Development and as an architect of the Presidential Climate Action Plan, a one-hundred-day action plan on climate that was presented to the Obama administration.

John A. Lanier joined the Ray C. Anderson Foundation as executive director in May 2013 to advance the legacy of Ray, his grandfather. He serves on the board of directors for Southface Energy Institute, the Southeast's nonprofit leader in the promotion of sustainable homes, workplaces, and communities through education, research, advocacy, and technical assistance. Previously,  Lanier was an associate attorney with Sutherland, Asbill & Brennan, LLP (now Eversheds Sutherland), specializing in US federal taxation. Lanier earned his juris doctorate from the University of Virginia School of Law, and he holds bachelor of arts degrees in history and economics from the University of Virginia.